Finding Your Way in the Undergraduate Physics Laboratory

Matthew M. J. French

Published by Pantaneto Press

Finding Your Way in the Undergraduate Physics Laboratory

ISBN 978-0-9549780-8-2

First published 2013

Printed and bound by CPI Group (UK) Ltd, Croydon, CR0 4YY

About the Author

Dr Matt French studied for an undergraduate physics degree at the University of Bristol. While working towards a PhD in experimental condensed matter physics, also at the University of Bristol, he spent 4 years demonstrating experiments in the first year undergraduate physics laboratory. Now an experienced school teacher and Head of Physics at Prior Park College, Bath he is well positioned to bridge the gap between students' knowledge at the end of school and that expected of them at the beginning of university.

Contents

List of Figures

List of Tables

1 Introduction

For some students, 'the experimentalists', their laboratory sessions are the highlight of their undergraduate degrees. It is in the laboratory that the physics they have learnt in lectures begins to make sense to them and the connections between the equations and reality are illuminated. Some of these students will continue working in laboratories to complete doctorates in experimental physics or turn to experimental work in industry. For other students, 'the theoreticians', the laboratory sessions are something that must be endured, often on a weekly basis. However, separating students into the discrete boxes of experimentalists and theoreticians does not really convey the holistic experience today's undergraduates receive: they must be talented at experiments as well as theory in order to excel at university.

This new book aims to guide both the 'experimentalist' and 'theoretician' through their compulsory laboratory course forming part of an undergraduate physics degree. Some students will have left school believing that physics experiments 'never work'. Designing and performing good physics experiments is a craft, it is a skill developed gradually over time. I hope this book might go some way to persuading all students of the value and beauty within a carefully planned and executed experiment and help them develop the skills to carry out experiments themselves.

Those students who are completing undergraduate research projects and beginning a doctorate will also find helpful and relevant sections. Graduate students supervising laboratory sessions will find this a useful resource: especially the dedicated section on 'Demonstrating Undergraduate Physics Laboratory'.

Chapter 2 includes the characteristics and safety considerations of a range of equipment and materials. Chapter 3 moves on to detail how many common measurement instruments are operated. In Chapter 4 the focus turns to how to analyse measurements and Chapter 5 deals with the presentation of results. Finally, Chapter 6 details a number of common experiments and example laboratory reports.

The idea of learning about experimental physics through reading a book, might at first seem to be tinged with a hint of irony. However, there are many concepts and techniques which a physics student needs to learn and understand. Often these are not specifically written down or taught in lectures: it is up to the interested student to investigate and learn for themselves. While the goal of this book is not to detract from this inquisitive and curiosity driven learning, it aims to provide an overview of the many topics and skills required in the first year undergraduate physics laboratory and beyond. Inevitably, some readers will find detail lacking in one area or another or pages which are currently superfluous to their needs. This book should not be read cover to cover, rather it should be dipped into to gain a basic understanding of a certain topic when necessary. Where needed the reader, then armed with a good grounding in the topic, can use other books or the internet for further research.

1.1 Characteristics of the Laboratory

In the first year students will usually work in pairs (or very small groups) undertaking a different experiment each week. It is important students are able to work with other scientists: for some experiments it is essential to have two people to perform the experiment. By discussing experimental methods and the problems experienced with their partner students can challenge their understanding and learn from each other. However, it is essential to work together as a team and that both are engaged in each part of the experiment. Each student should record their own results and perform their own analysis. Of course, the calculations and results after each stage in the analysis should be cross checked. Identifying each other's mistakes soon after they are made will work to the advantage of both students.

Laboratory sessions will probably consist of a day, or most of a day, to perform the experiment and collect results. Students may have to

Figure 1: The undergraduate laboratories at the University of Oxford. Left: condensed matter laboratory. Right: optics laboratory.

Figure 2: The undergraduate laboratories at the University of Bristol.

hand in their analysis and write up of the experiment the same day, or be given a number of days to complete this. Usually a number of the experiments are written up as formal laboratory reports, this is discussed in more detail in Section 5.2.

When entering the laboratory, the experiments will usually be partially set up on tables in a large room: see for example Figure 1 and Figure 2. There will probably be other students nearby performing the same experiment using an identical set of apparatus. This provides more opportunities to discuss results and troubleshoot the equipment and the method when results don't seem to make sense. Each group of 8 or 10 students (4 or 5 pairs) will usually be supervised by a post graduate laboratory demonstrator. When advice or help is needed, having carefully thought about the question and tried to find the solution the student should consult the demonstrator for help. Sometimes, the demonstrator will remain with the same group of students for the whole year working with them each week on different experiments. In other cases a demonstrator will remain with the experiment and work with a different group of students each week.

Never touch or remove equipment from someone else's experiment. Experiments can be working and recording data even if the user is not present. It is very annoying and frustrating if an experiment is carefully setup, recording data and someone disturbs or removes some of the equipment.

In the second year, and perhaps third year laboratory courses experiments become more complex and longer: perhaps 4, 6, or 10 days (usually at 1 day per week). Students still work in pairs, but the scripts become less detailed and it is expected that students use the knowledge and skills acquired in the first year laboratory to develop a method and derive much of the analysis themselves.

1.2 Demonstrating Undergraduate Physics Laboratory

This section is written as a brief introduction for anyone demonstrating undergraduate physics laboratory: usually this would be concurrent with studying for a PhD.

Demonstrating undergraduate physics laboratory can be an enjoyable experience which also provides a little extra money. It is also good experience for the future when applying for post-doc jobs or academic posts. It is usually expected that all academics working at a university contribute to teaching in some way. When spending much of the time working on the computer reading papers or analysing data, a day in the undergraduate laboratory can give a very welcome break. At other times, when experiments are going well and significant progress is being made the interruption caused by the laboratory day can be a little frustrating.

In some universities or year groups the demonstrator is attached to a group of around 8 – 10 students who work in pairs. Each week a different experiment is performed and it is the demonstrators' responsibility to ensure they know how to carry out the experiment according the the script given to the students, use all the necessary equipment and complete the analysis. Demonstrators should be on hand with their groups of students for the full duration of the laboratory session. At least initially, this can seem like a great deal of work and a little daunting for anyone not confident with the experiments. However, there is usually the opportunity to practise the experiments when the laboratory is closed to undergraduate students and more experienced laboratory demonstrators who have worked for a number of years are usually happy to help.

Alternatively, in some universities or year groups the demonstrator is attached to a particular experiment (or a small number of experiments). This allows the demonstrator to become an expert in certain experiments, although this means they meet different students each week.

The laboratory books of the students are usually handed to the demonstrator for checking every week. The demonstrator will be responsible for marking the work and submitting the marks for collation. It is good practice for each demonstrator to keep a record of all marks they award. It is wise to always keep on top of marking. It need not take significant amounts of time, especially if it is done as soon as possible after the books are handed when it will be more fresh in the mind. Calculations and error analysis need not be worked through for each student individually and checked in detail, although sometimes it is easy to spot and highlight/correct an error the student has made.

Usually some guidance is provided to help with the assignment of marks, but the following could be used as a guide if nothing is provided. Bear in mind that a 2:2 would correspond to 5/10, a 2:1 to 6/10 and a 1st to 7/10.

A mark of 4/10 or less would be awarded for a write up which did not meet the criteria for a mark of 5/10. This will generally be awarded rarely.

A mark of 5/10 would correspond to a reasonable attempt to collect data in the laboratory, perhaps incomplete due to a lack of effort, skill/patience or hindered by temperamental equipment. There should be an attempt to describe the method in some detail and to complete the analysis as far as a final answer. The final answer may be significantly incorrect due to errors in the recoding of data or the analysis. There may be no consideration of uncertainty.

A mark of 6/10 would be awarded where there was a complete or almost complete set of data, perhaps only limited by temperamental equipment. The majority of the steps in the method are described in the correct order. The analysis has been completed and a final answer has been calculated. The final answer should be order of magnitude correct, or there are one or two simple errors which can be easily identified such as a missing constant, an incorrect conversion or a power of ten error. There may some consideration of uncertainty in the final result but a reasonable, quantitative estimate is not given.

A mark of 7/10 would correspond to a collection of a complete set of data of an appropriate accuracy and precision. All the key points in the method would be described in the correct order and in significant detail. The analysis should be complete through to the final answer. The final answer should be order of magnitude correct or any larger discrepancy identified and an explanation attempted. There is some consideration of uncertainty leading to a quantitative estimate of the uncertainty in the final answer, although its origin (both in the experiment and in numerical terms) may be somewhat unclear.

A mark of 8/10 or higher will probably only be awarded rarely. This would correspond to meeting and exceeding all the criteria for a mark of 7/10. There should be clearly recorded data which has been repeated a number of times (as time allowed) along with a detailed consideration of the uncertainty in each measurement. The method should provide significant detail, which would allow a similar student to perform the same experiment without reference to other texts. The final answer should be quoted with an uncertainty which is clearly calculated and well founded in the actual measurements taken.

Alternative guidance is sometimes provided for marking formal reports. Marks might be awarded in categories such as 'Clarity of arguments', 'Scientific Content', 'Presentation', 'Scientific Literacy and Style'.

The 'Clarity of arguments' section might score 7/10 if there is a clear introduction, explanation of the method, the analysis and the results. This should be at a level understandable by a student in the same academic year as the author. If some of these areas are lacking a clear explanation then 6/10 may be more appropriate. If most of these areas are not clearly explained then 5/10 or less may be appropriate.

The 'Scientific Content' section might score 7/10 if there is sufficient detail on all aspects of any derivations needed, the method, analysis and error analysis which would allow a similar student to repeat the work without reference to other texts. Extremely thorough and competent work would score $\geq$8/10, and work where some sections are not fully addressed would score 6/10. If many sections are not fully addressed or if any are missing a mark of $\leq$5/10 would be appropriate.

In the 'Presentation' section marks are awarded for the layout and organisation of the work. Tables and diagrams should be neat, appropriate and fully labeled with column headings, units and figure captions (eg. Figure 2: A diagram of ...). Graphs should be of an appropriate type (usually scatter graphs, without the points joined up) have axis labels, units, a line of best fit (in most cases), a figure caption and be large enough to be seen clearly. The written work should be logically ordered and there should be a good use of section numbers and sub-headings. Where there are only a few minor mistakes a mark of 7/10 should be awarded. A mark of 6/10 would be appropriate if there are a number of mistakes or one point is consistently absent. If there are many of the above points absent then a mark of $\leq$5/10 could be awarded.

Finally marks in the 'Scientific Literacy and Style' are awarded for the style in which the report is written. Usually reports should be in a formal style with no use of the words 'we', 'I' or 'you'. Description of the physics involved and the analysis should make correct use of appropriate technical terms. Reference should be made to other work which is cited in an appropriate, formal style. If these criteria are mostly met then a mark of 7/10 should be awarded. If there are occasional mistakes, then 6/10 could be given. With many mistakes or a whole report written in the first person (using 'we','I' or 'you') then a mark of 5/10 could be awarded.

2 Safety

While performing experiments at school students probably took it for granted that the equipment they used was safe and was very unlikely to do them harm unless used in an obviously unsafe manner. The class teacher would have been following carefully thought out advice from a national organisation called CLEAPSS [1], in addition to performing their own risk assessment. In most cases common sense and following the teachers' instructions was all that was required on the student's part.

At university, students are likely to meet liquid gases, dangerous chemicals, high temperatures, high (electric) potential differences and lasers. Risk assessments should have been carried out for all instructed experiments met in the first two or three years of an undergraduate degree. However, some of the equipment or chemicals used will have the potential to cause serious harm. Furthermore, other equipment or chemicals which may be present in the same area could also be very dangerous.

Students should make sure risk assessments are read and understood and follow any safety rules or advice given by the university: students have legal obligation to do this in the United Kingdom. Students should begin to see why certain precautions are needed and develop responsibility for themselves. Equipment or chemicals which students do not understand or are not confident with should never be touched or used. Where one person may be confident, another may not be: students should always ask for advice if they are unsure or not confident with certain equipment.

There is general safety advice which is applicable to all laboratories at all time. Later, safety advice for specific pieces of equipment will be covered.

Shoes should always be worn when working in the lab because they provide some protection from spillages and falling objects. Sandals, high heels and open shoes are completely inappropriate as they allow chemical or cryogenic liquids (used by the students themselves or others) to pose a danger to the feet.

Loose or dangling clothing, jewelery, hair and beards may get caught in moving equipment: especially motors. Sensible clothing should be worn, jewelery removed and hair tied back using a hair band or elastic band.

If working with chemicals, clothing should ideally cover legs and arms and a lab coat should be worm to help protect against spills or splashes.

Food and drinks should never be present or consumed in the lab. There is the risk of contamination of the food from chemicals, including traces of unknown chemicals from previous work which took place in the same area. Drinks can easily be spilled causing damage to expensive equipment.

2.1 Hazard Symbols

The Globally Harmonized System of Classification and Labeling of Chemicals (GHS) uses the pictograms shown in Table 1. There is not officially a single word meaning for any of the symbols: there should be adjacent written hazard information. However there is a general guide given here based on the old European descriptions of similar hazard symbols. Further non-chemical hazard symbols which might be present in a laboratory are shown in Table 2. The symbols may be shown with or without the accompanying text.

2.2 Electrical

All devices provided in the lab powered by mains electricity should have been tested prior to use by the university to ensure they are safe. This is called a portable appliance test (PAT) and is carried out at set time intervals depending on the equipment. It involves a visual examination of the wire and plug and a resistance measurement to ensure that there is a good earth connection from any exposed metal

Explosive	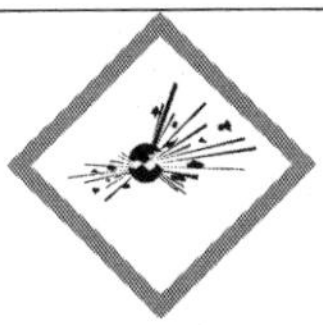
Corrosive	
Longer term health hazards such as carcinogenicity and respiratory sensitisation	
Flammable	
Less serious health hazards such as skin irritancy/sensitisation	
Contains gas under pressure	
Harmful to aquatic life	
Toxic	

Table 1: Chemical Hazard Symbols. Usually the border is red and the diagram is black.

Table 2: Non-Chemical Hazard Symbols. Usually the text, border and diagram is black and the background is yellow.

on the case. Because it was deemed safe at the time of testing, does not guarantee it is safe at the time of use.

Before plugging in any device to the mains it is prudent to visually check the cable and plug for any cracks, cuts, breaks or exposed wires. Any equipment which is found damaged or not working should not be used. It should be reported to a lab demonstrator or lab technician who will remove the equipment or label it as unserviceable. Never use equipment marked as unserviceable or U/S or which has had the mains plug cut off.

Many devices operate from mains power or have a built in transformer and/or AC-DC converter. Additionally mains powered, power supplies may be used. There are two major categories of power supplies which adapt the standard mains supply (UK and Europe: 230 V at 50 Hz and USA: 115 V 60 Hz) to power other devices: low emf power supplies or extra high tension (EHT) supplies.

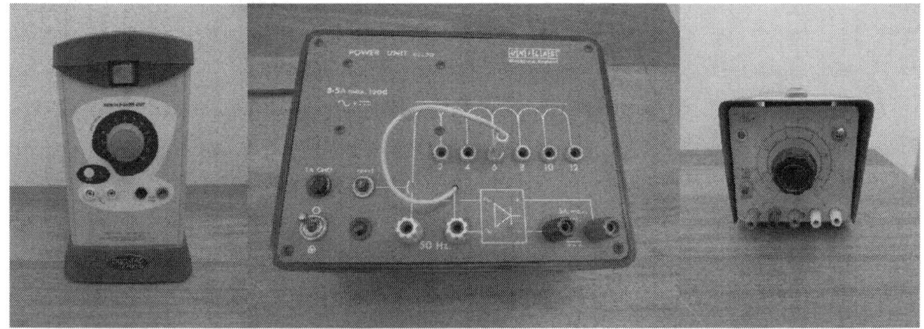

Figure 3: Examples of simple low emf power supplies. Usually the yellow outputs give AC. The black (negative) and red (positive) outputs give DC.

Devices which require a particularly large amount of power may be connected to a special 'three-phase' socket.

Low emf power supplies are classed as those supplying up to 28 V AC or 40 V DC. At this (electric) potential difference, exposed connectors can be used as this (electric) potential difference is generally classed as safe to touch. Examples of low emf power supplies are shown in Figure 3 and suitable exposed connectors are shown in Figure 4. Hand built digital electronic circuits and most apparatus will comfortably fall within the limit of a low emf power supply. Working with emfs above these limits should be avoided if at all possible.

2.3 High Voltages

Where there is a scientific need to use (electric) potential differences higher than 28 V AC or 40 V DC, great care must be taken to avoid anyone touching exposed contacts. If the device is mains powered, this should have been built into the design of the device. Sometimes an EHT power supply will be needed which may deliver in the region of 5000 V (5 kV) or more. Special connection cables should be used which have shielded connectors to reduce the risk of electrocution: see Figure 4.

Figure 4: Left: Low (electric) potential difference connectors with exposed metal contacts. Right: High (electric) potential difference connectors with shielded metal contacts. The shielding automatically moves back as the plug is connected into a socket. Both leads can be piggy-backed to connect multiple leads to the same place.

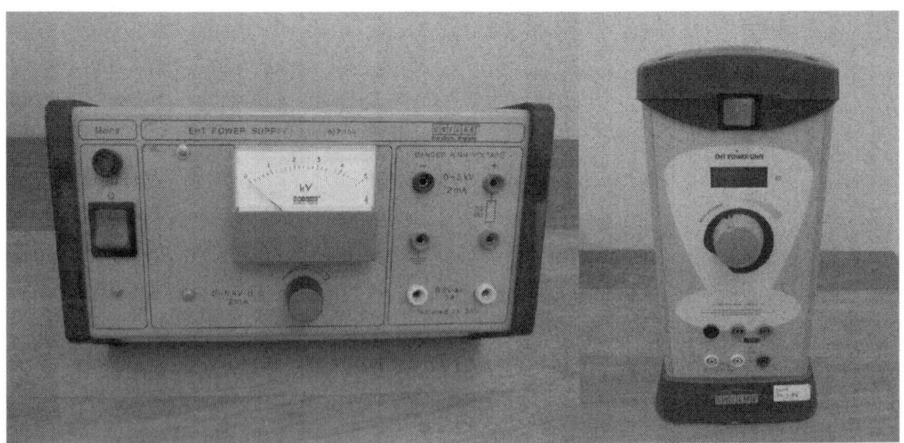

Figure 5: Examples of high emf (EHT) power supplies.

No matter what form of power supply is connected to a device it is always safest to turn off the supply before touching any metal or 'live' component. Special care should be taken with EHT power supplies: they must not simply be switched off using a power switch (on the power supply or plug socket). The (electric) potential difference should be slowly decreased to zero and a voltmeter (or voltage display built in to the power supply) should be used to ensure the potential difference has reached zero before the power supply is switched off. The risk of switching off an EHT incorrectly is the possibility that circuit elements or components (most likely a capacitor) may still retain significant separated charge. This charge creates a high potential difference which can cause a dangerously large current to flow through you when any exposed wiring is touched (even if the power supply is switched off and unplugged from the mains).

2.4 Gas Cylinders

Cylinders of gas provide a reliable and consistent supply of pure, dry gas to an experiment. Common gases used in university experiments include oxygen, carbon dioxide, nitrogen, helium, argon and hydrogen.

The use and movement of gas cylinders can be potentially dangerous, although it can be done safely if the risks and dangers are understood.

Gas cylinders should always be stored and used in an upright position. They must always be secured so as they can't fall over: usually in a transport trolley or using a chain or belt which securely attaches the cylinder to a wall or table bracket: see Figure 6 and 7. Cylinders which are not secured may fall over and cause a crushing injury to someone nearby. A further severe danger is that the regulator may be sheared off, allowing gas to escape rapidly turning the cylinder into a torpedo capable of major injury and penetrating concrete walls [2].

Generally, cylinders not in use will be stored outside in a cage to keep them secure. Great care should be taken if a gas cylinder must

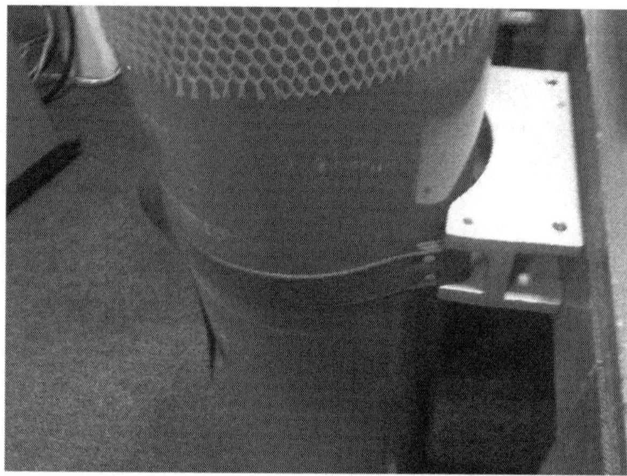

Figure 6: A bracket and belt anchoring a cylinder to the desk.

be moved. Cylinders are heavy and have no 'carrying handles'. They should be strapped securely into a transport trolley: see Figure 7. Care should be taken when moving trollies as they usually need to tipped and part of the weight of the cylinder supported by the user's arms. Trolleys should be taken slowly and carefully over uneven floors and doors should be propped open in advance.

Gas cylinders store gas under high pressure, commonly this is around 230 bar. To reduce this pressure for use in an experiment a regulator must be used. The regulator is attached to the outlet at the top of the cylinder. Regulators for hydrogen and other flammable fuel gases have a left handed thread on the nut, whereas regulators for oxygen and inert gases have a right handed thread. Special care should be taken to ensure that regulators for oxygen cylinders are labeled as oil free: usually with a crossed out symbol of an oil can. This reduces the risk of fire.

To fit a regulator to a cylinder begin by ensuring the regulator is turned off by turning the regulator knob anticlockwise. Next check the screw threads on the cylinder and regulator are in good condition. With closed eyes, blow into the cylinder head to remove any dust. Then

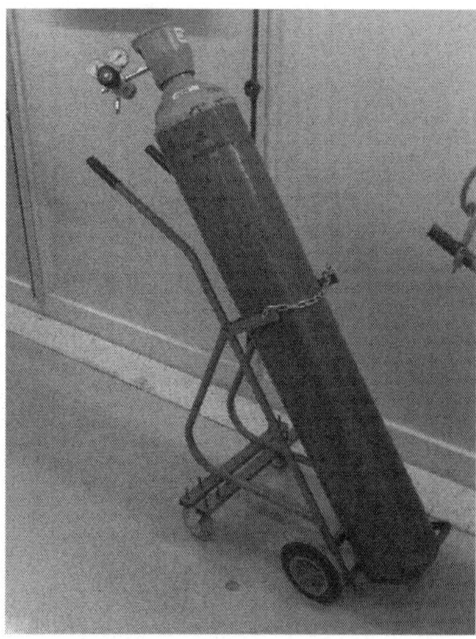

Figure 7: A trolley used to move gas cylinder. This trolley has an extra set of wheels so the user doesn't have to support part of the weight of the cylinder while moving the cylinder.

place the regulator above the thread and using a spanner, turn the nut (clockwise from above for right handed threads) until the nut will not turn any further. Do not use excessive force.

To obtain gas, ensure the knob is turned fully anticlockwise (the off position). Use the cylinder key to turn the spindle (see Figure 9) fully anticlockwise (the on position). The pressure gauge now gives a reading, indicating the contents of the cylinder. As gas is used the pressure reading will decrease: an empty cylinder will read approximately zero when the spindle is open. A quick check for leaks in the regulator can be made. Gas leaking rapidly will make a hissing noise. Slowly turn the regulator knob clockwise (to the on position) to obtain the gas. The second gauge indicates the delivery pressure of the gas. The further the knob is turned clockwise the higher the delivery pressure.

To close down a gas cylinder when the experiment is finished turn the regulator knob fully anticlockwise to the off position. Use the cylinder key to move the spindle fully clockwise to the off position.

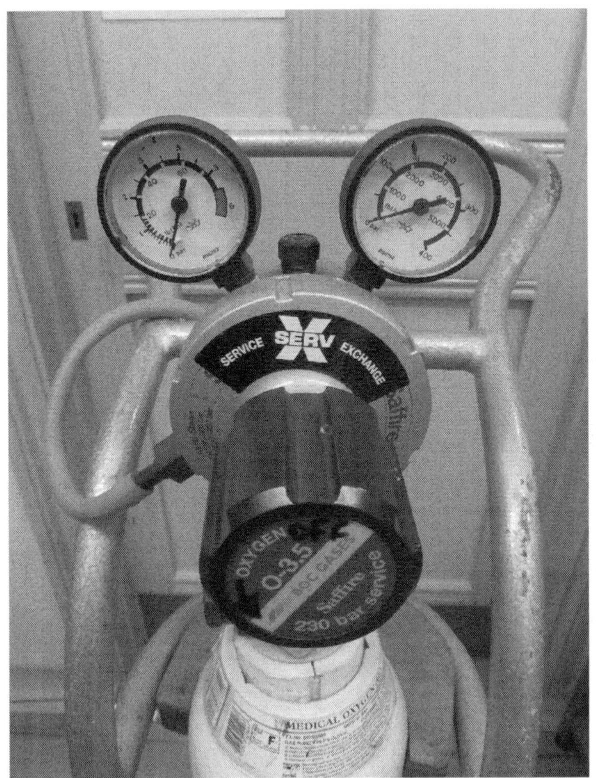

Figure 8: A gas regulator. This one is for oxygen only: the knob is labeled and the crossed out symbol of an oil can is present in the pressure gauges.

Figure 9: Left: A gas cylinder key. Right: The spindle on a gas bottle.

Next turn the knob clockwise to remove any gas in the regulator. This prevents damage to the diaphragm in the regulator. The pressure gauge will now read zero. Finally turn the knob fully anticlockwise again.

To remove the regulator when the cylinder is empty and needs changing follow the instructions in the paragraph above to close down a gas cylinder. Then unscrew (anticlockwise from above for a right handed thread) the nut using a spanner. The regulator should now be free. It is vital that the spindle is fully turned off (turned fully clockwise) before starting to unscrew the nut. This prevents gas escaping rapidly from the cylinder potentially causing the torpedo effect described earlier.

Note carefully that the spindle is off when turned fully clockwise where as the regulator is off when turned fully anticlockwise.

Compressed gases are graded by their purity which is usually expressed as a percentage, for example: 99 % or 99.999 %. However this is often re-expressed in terms of the number of nines: 99 % becomes 2N (corresponding to 2 nines) and 99.999 % becomes 5N (corresponding to 5 nines). A further variation is 99.95 % which becomes 3N5 (corresponding to 3 nines and then a five).

In the EU, the colours of cylinders of some gases are becoming standardised following the new standard EN 1089-3. Colours for cylinders of the most common gases are given in Table 3.

2.5 Vacuum Pumps

Vacuum pumps are used to remove air or other gases from a sealed container. Depending on the pressure required different types of pump are used. Commonly used pumps include a simple wine-saver hand pump, a rotary pump, a diaphragm pump, a diffusion pump or a turbo pump.

The primary danger when using a vacuum pump to reduce the pressure inside a container is that the container walls fail and the container implodes. Wall pieces may then rebound off each other and fly out-

Gas	Cylinder Colour
Argon	Green
Carbon Dioxide (gas withdrawal)	Black, Grey top
Carbon Dioxide (liquid withdrawal)	Black, Grey top, vertical White line
Helium	Brown
Hydrogen	Red
Nitrogen	Grey, Black top
Oxygen	Black, White top

Table 3: Compressed gas cylinder colours.

wards causing injury. It is usual for glass or plastic pressure vessels (such as bell jars) to be surrounded by a mesh cage or a safety screen to provide some protection in the event of an implosion.

The magnitude of the force acting on a container under vacuum can be demonstrated with a used soft drinks can. A small amount of water is placed in the bottom of the empty can which is then heated. Once the water is boiling and the can contains steam, the can is then rapidly turned upside down with the opening submerged in water to seal the can. The water rapidly cools and the steam condenses reducing the pressure in the can causing it to fold inwards. A reduction of pressure of around 30 mbar (to around 970 mbar) in the can is sufficient to cause it to implode.

2.5.1 Pressure Units

Most pumps or vacuum systems will have a pressure gauge to measure the pressure achieved. The gauge may read in a variety of units so it is important to be aware of how to inter-convert units. The SI unit of pressure is the $\mathrm{N/m^2}$ derived from the equation $P = F/A$. $1\,\mathrm{N/m^2}$ is identical to 1 Pascal, abbreviated to 1 Pa. Other common units for pressure include the atmosphere (atm), the Torr (which is equivalent to mm of Mercury) and the pound per square inch (psi). Conversion

	1 bar	1 atm	1 Pa	1 Torr	1 psi
bar	1	1.01325	1×10^{-5}	1.33322×10^{-3}	0.068948
atm	0.98692	1	9.8692×10^{-6}	1.31579×10^{-3}	0.068046
Pa	100000	101325	1	133.323	6894.8
Torr	750.06	760	7.5006×10^{-3}	1	51.7149
psi	14.5038	14.6960	1.45038×10^{-4}	0.0193368	1

Table 4: Conversion factors for common pressure units.

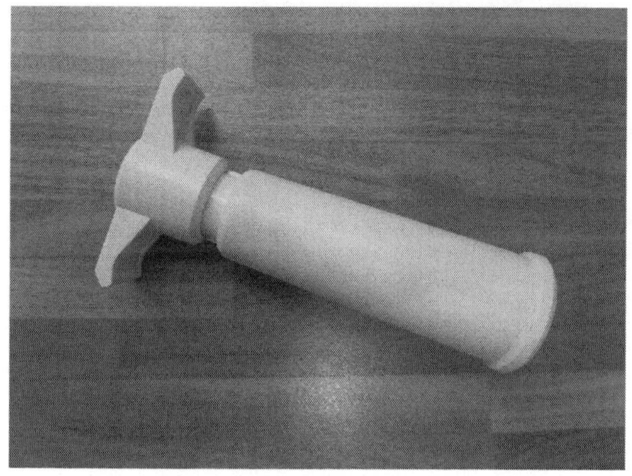

Figure 10: A vacu-vin hand pump.

factors between these units are shown in Table 4.

2.5.2 Hand Pumps and Vacu-vins

A hand pump or vacu-vin is primarily designed for removing air from bottles of wine, thus preserving an opened bottle of wine for longer (see Figure 10). When the appropriate pressure is reached this is indicated by the device with a 'click'. Only rarely do these have a built in pressure gauge, but typically they will achieve a pressure of around 400 mbar before the 'click'. Inside the vacu-vin a piston compresses air from the vessel being evacuated and forces it out through a valve.

Figure 11: A rotary pump.

2.5.3 Diaphragm Pump

Diaphragm pumps consist of a vibrating diaphragm which compresses air from the experiment and forces it out through a valve. Typically they achieve a pressure of around 20 mbar.

2.5.4 Rotary Pump

Rotary pumps are the work horse pump of the physics lab. They provide a cheap, reliable way to achieve a vacuum down to a pressure of around 0.1 mbar. A two stage rotary pump consists of two rotary pumps in series and will produce a vacuum of around 0.001 mbar = 1×10^{-3} mbar. Figure 11 shows a rotary pump.

The pump consists of a vane projecting from a cylinder rotating eccentrically which compresses air and forces it out through a valve. Rotary pumps have oil inside. This is a special type of oil which has a very low vapour pressure i.e. it produces very little vapour. If it did not have a low vapour pressure it could not be used as the oil vapour would pollute the vacuum the pump was trying to produce. Despite the low vapour pressure tiny droplets (i.e. a mist) of oil from the pump can be carried away by the air it exhausts. This is more of a concern when the pumping flow rate is high or the pump is run for prolonged periods of time or in a confined space/small room. There are statutory limits on the amount of oil mist permissable in air and so rotary pumps are usually fitted with an oil mist filter.

Some pumps have an anti-suck provision built in. If not, oil will be sucked from inside the vacuum pump back into the experiment. To prevent this the experiment must be isolated from the pump by closing a valve before the pump is switched off. Air should then be allowed into the pump using the air inlet valve. The experiment can be brought back to atmospheric pressure when needed using a separate air inlet valve in the experiment: if the air inlet valve on the pump is used oil will be sucked into the experiment.

2.5.5 Diffusion Pump

In a diffusion pump molecules of gas are swept away by jets of oil vapour. The exhaust must be connected to a rotary pump: that is, a diffusion pump must be backed by a rotary pump. Diffusion pumps have a high displacement rate even at low pressures and so will reach low pressures despite minor leaks. They can achieve a pressure of around 1×10^{-4} mbar.

2.5.6 Turbomolecular Pump

A turbomolecular pump (see Figure 12) contains a rapidly spinning turbine which repeatedly collides with gas molecules imparting momentum, directing the molecules towards the exhaust. The exhaust must be connected to a rotary pump: that is a turbomolecular pump must be backed by a rotary pump and can achieve a minimum pressure of around 1×10^{-8} mbar.

To prevent the turbine blades becoming damaged due to too many collisions with gas molecules, turbomolecular pumps should not be

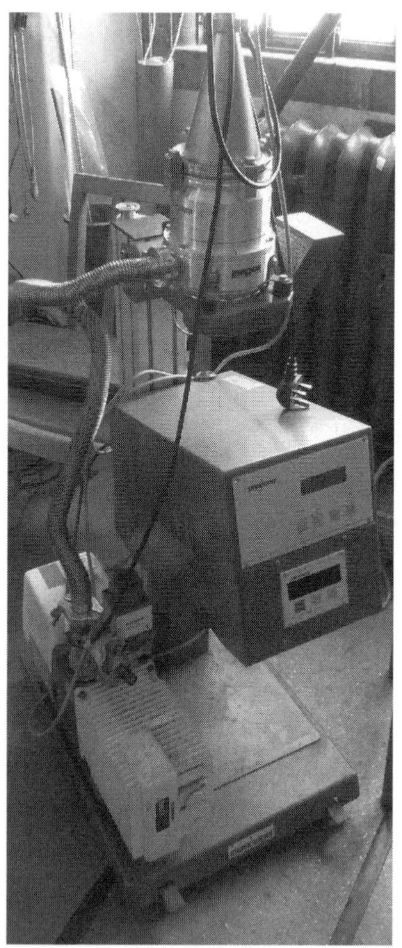

Figure 12: A turbomolecular pump backed with a rotary pump.

turned on until the rotary backing pump has achieved a pressure of at least 9×10^{-3} mbar. Care must be taken when switching off a turbo-molecular pump. Air should not admitted to the system until the turbo pump has slowed to $25 - 50\,\%$ of its maximum revolutions per minute. This may take a considerable time of the order of $15 - 30$ minutes. Air should then be admitted very slowly through the special air inlet valve.

2.6 Cryogenics

Cryogenics refers to the production of very low temperatures in the laboratory. Rather than using a refrigerator such as you might find in a kitchen, in the physics lab liquid cryogens are generally preferred. This is because of their simplicity, relatively low cost and the complex nature of cryogen free coolers.

Common liquid cryogens are nitrogen and helium. These elements are gases at room temperature and pressure, however can be cooled and liquified. Often cryogens are produced off site by a specialist company and purchased by a university. Liquid cryogens provide a simple reliable method to cool other objects to very low and stable temperatures. It is likely that liquid nitrogen and dry ice will feature early on in many undergraduate courses and so this section begins by discussing how these can be used safely. Liquid helium will also feature in some undergraduate laboratories and so this will be mentioned briefly.

No written publication or instructions can never be a substitute for proper training by a university before any cryogenic liquids are used. This ensures that the student is aware of all the dangers specific to the equipment and setting in which they will be using cryogens.

2.6.1 Liquid Nitrogen

At atmospheric pressure nitrogen boils at 77 K (-196°C) and it has a latent heat of vapourisation of 200 kJ/kg. In its liquid form it is a clear, colourless liquid. Nitrogen freezes creating nitrogen ice at 63 K (-210°C). By reducing the pressure of the gas above a container of liquid

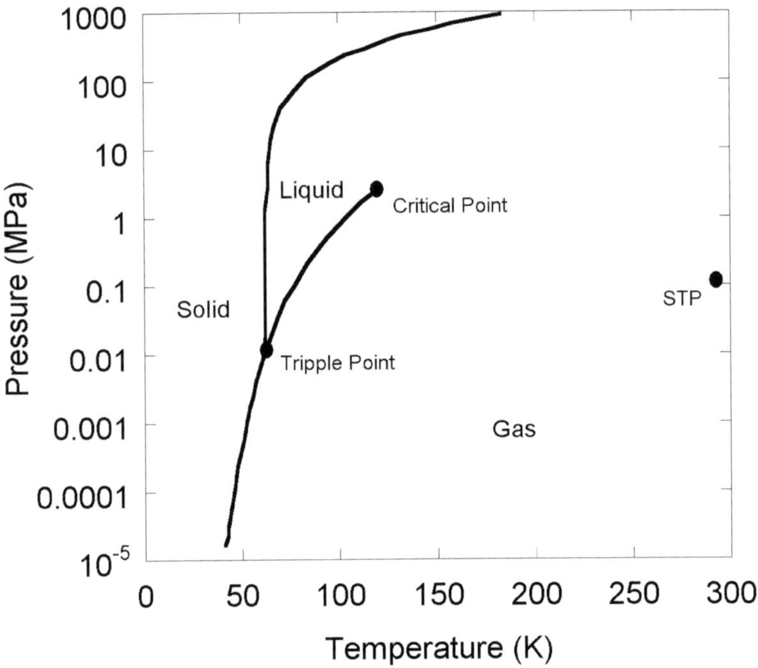

Figure 13: Phase diagram for Nitrogen. Nitrogen is a gas at STP (standard temperature and pressure).

nitrogen with a rotary vacuum pump it is possible to freeze nitrogen. Figure 13 shows the phase diagram for Nitrogen.

Due to its high latent heat of vapourisation and relatively low cost it is commonly used to cool objects in the physics laboratory. It will be stored in a Dewar flask (named for the Scottish scientist James Dewar 1842 – 1923) which is essentially a large vacuum flask of the type used for hot drinks: see Figure 15. If a vacuum flask intended for food use is used it is vital the screw top is not used as this will lead to a build up of pressure as the nitrogen evaporates and a possible explosion. For easy access to small quantities it may be temporarily stored in a polystyrene or neoprene bucket: see Figure 14.

It is reasonably safe to use in the laboratory provided there is some understanding of the dangers it poses and simple safety procedures are

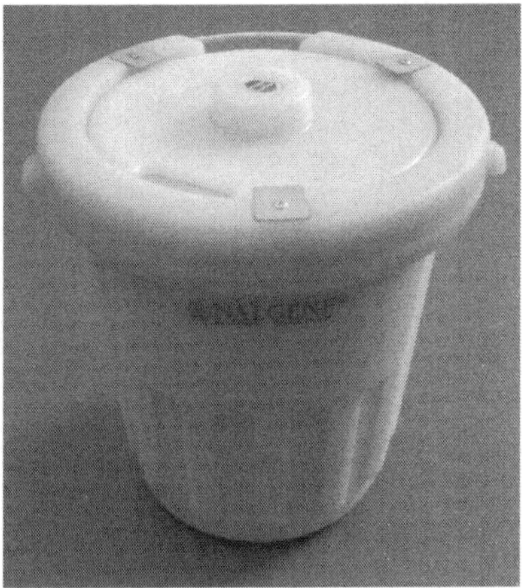

Figure 14: A neoprene plastic bucket for storing liquid nitrogen.

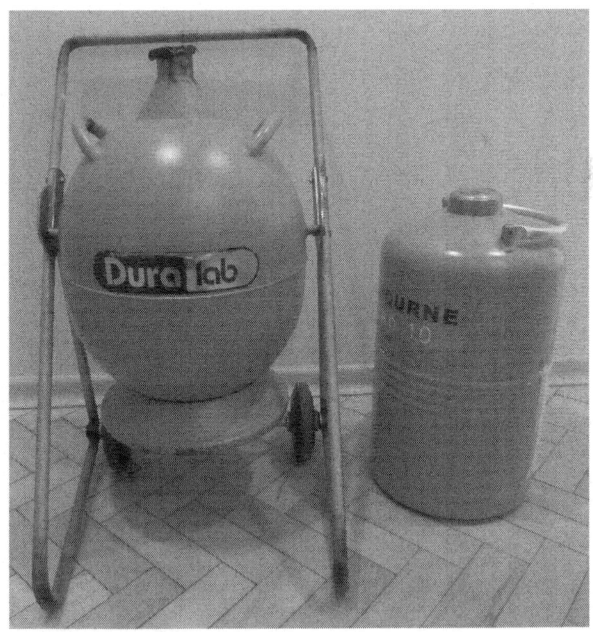

Figure 15: Liquid Nitrogen Dewars. Left: a 25 litre 'onion' Dewar with a trolley Right: a 10 litre Dewar.

followed.

The primary danger is the cooling and freezing of parts of the human body. However, if liquid nitrogen comes into gentle contact with skin it rapidly evaporates forming a layer of nitrogen gas which protects the body from any further contact with the liquid as well as insulating the body from the coldness of the liquid. This is called the Leidenfrost effect and can also be seen if water is spilt on a hot electric cooker plate. Thus it is perfectly safe to dip fingers into liquid nitrogen for a short time (less than 1 second) while being careful to avoid touching any container holding the nitrogen. It is also possible to pour liquid nitrogen over hands or other parts of the body provided it is not poured from a significant height nor allowed to collect in the cup of hands, pockets, sleeves, cuffs or shoes. Liquid which collects in clothing will be in contact with the body for a long period of time and could lead to severe cold burns.

This possibility of collection of liquid absolutely rules out the use of gloves when handling liquid nitrogen. Vitally however, solid objects, especially metal, cooled by liquid nitrogen should never be touched with bare hands: thick gloves (suede gardening gloves or special cryogenic gloves may be appropriate) must always be worn. Bare hands will stick to cold objects (rather than follow the recoil reaction experienced when hot objects are touched) and present the danger severe cold burns.

Care should be taken to avoid splashes of liquid nitrogen entering the eyes especially when pouring liquid nitrogen from one container to another and so wearing goggles may be appropriate. Open shoes or sandals must be avoided by all those working in the same room as liquid nitrogen is used.

A further danger presented by liquid nitrogen is asphyxiation. When a liquid turns into a gas it expands significantly (remember the diagrams showing the particles in a solid, liquid, gas drawn at school). In the case of liquid nitrogen warming to room temperature the expansion factor is 683 (i.e. 1 liquid litre expands to 683 gas litres at room temperature and pressure). Therefore if even a small amount of liquid nitrogen is spilt in a small room, lift, car or other confined space it can displace the oxygen causing asphyxiation. This danger can be avoided by never traveling in a car or a lift with liquid nitrogen and ensuring that the quantity of liquid is always appropriate for the room in which it is being stored or used.

This paragraph contains a detailed calculation regarding the minimum room volume in which a 25 litre Dewar of liquid nitrogen could safely be stored or used assuming there is no low oxygen alarm linked to a fail safe forced ventilation system. The BCGA Code of Practice [3] states that "*From a normal oxygen level of 21 %, a room shall be sufficiently ventilated for the two, normal conditions not to cause a reduction in oxygen concentration below 19.5 %: (i) the normal evaporation of all dewars and liquid nitrogen containers within the room (ii) the filling losses from filling the largest dewar from a warm condition. Additionally, the complete spillage of the contents of the largest dewar shall not cause the oxygen concentration to fall below 18 %*". Following Annex 3 of [3], accounting for the normal evaporation rate of the 25 litre Dewar the concentration of oxygen in the room after a long period will be

$$C = \frac{V_r \times 0.21 \times n}{L + (V_r \times n)} \tag{1}$$

where $V_r$is the room volume, L is the gas release units of m^3/h and n is the number of air changes in the room per hour. Assuming a typical value of $n = 0.4$ and that L is twice the manufacturers quoted value of 0.2 liquid litres per day (i.e. $L = 0.4 \times 683 = 273$ gas litres per day $= 11.4$ gas litres per hour $= 0.0114\ \mathrm{m}^3/\mathrm{h}$ where the gas expansion factor is 683) gives the minimum room volume to not fall below 19.5 % oxygen as $0.38\ \mathrm{m}^3$. This is insignificant compared to the requirement concerning the complete spillage of a Dewar. Following Annex 4 of the same code, accounting for the complete spillage of the 25 litre dewar the minimum room volume for storage/use will be given by

$$C = \frac{0.21\left(V_r - \frac{V_d \times f_g}{1000}\right)}{V_r} \tag{2}$$

where C is the oxygen concentration, V_d is the dewar capacity in litres and f_g is the gas expansion factor. Solving this leads to a minimum room volume of 120 m^3 that the oxygen concentration will not fall below 18 %. This should be appropriately adjusted if more than 1 Dewar is present.

Where asphyxia is possible (i.e. where liquid nitrogen has been used or spilled in a confined space) if any of the following symptoms appear immediately remove the affected person to the open air, and follow up with artificial respiration if necessary: rapid and gasping breathing, rapid fatigue, nausea, vomiting, collapse or incapacity to move or unusual behaviour.

First aid for cold burns should include running the affected area under warm (but not hot water), covering with clean cling film and a prompt visit to the accident and emergency department at a hospital.

2.6.2 Dry Ice

Dry ice is the name given to solid (frozen) carbon dioxide because it sublimes (i.e. turns directly from a solid to a gas with no liquid phase) as its temperature increases above 194.5 K (-78.5°C). This can be seen by studying the phase diagram in Figure 16. Its density when solid at 194.5 K and 1 atmosphere is 1560 kg/m^3 and when a gas at 273 K and 1 atmosphere is 1.98 kg/m^3. This is more dense than air (1.3 kg/m^3) so it sinks to the bottom of a container or a room and any leaks large releases of gas will fill the room from the floor upwards.

Dry ice is usually supplied in the form of white cylindrical pellets around 8 mm in diameter and 1 – 2 cm in length. It can usually stored in a thick polystyrene box with a loosely fitting lid for a few days.

The primary danger from handling dry ice is severe frost bite, there-

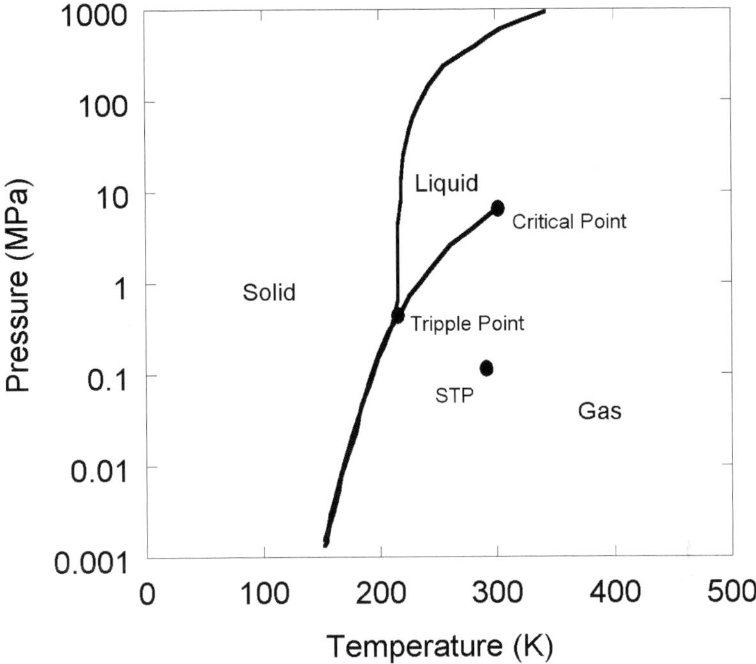

Figure 16: Phase diagram for Carbon Dioxide. Carbon Dioxide is a gas at STP (standard temperature and pressure).

fore dry ice should be handled only while wearing thick gloves (suede gardening gloves or special cryogenic gloves may be appropriate). Similar gloves should also be used while handing objects cooled by dry ice. With extreme care, individual pieces of dry ice my be handled without gloves provided they are not gripped tightly nor held for more that a second or two.

As with liquid nitrogen, care must be taken over storage especially in small room or confined spaces. The typical carbon dioxide content of the atmosphere is around 390 ppm (parts per million) [4]. The safe limit is considered 5000 ppm, above this level symptoms consist of an increased breathing rate, tiredness and headaches. If any of the symptoms appear immediately remove the affected person to the open air and follow up with artificial respiration if necessary.

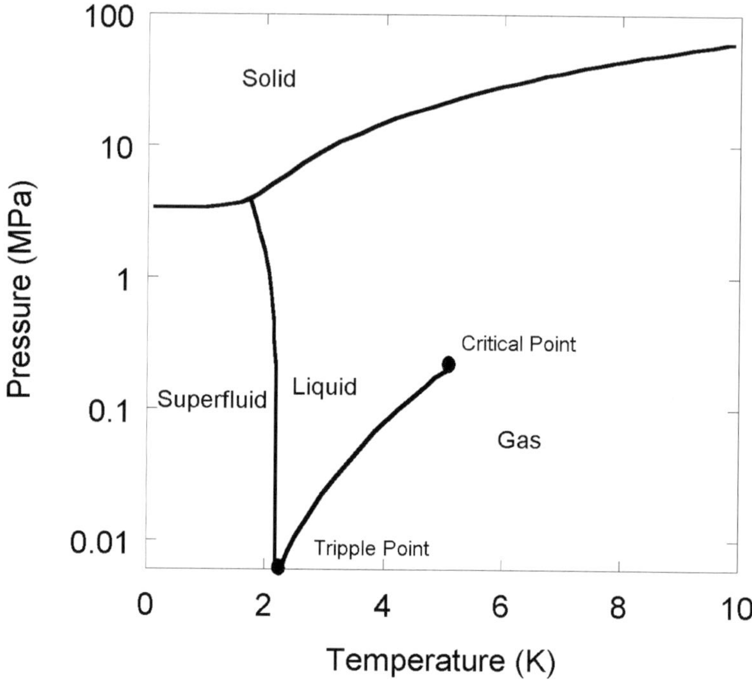

Figure 17: Phase diagram for Helium-4. Helium-4 is a gas at STP (standard temperature and pressure).

2.6.3 Liquid Helium

At atmospheric pressure liquid helium boils at 4.2 K (-269°C) and it has a latent heat of vapourisation of 21 kJ/kg. Helium does not solidify at atmospheric pressure. As a gas it is less dense than air with a density of 0.179 kg/m^3 at 273 K and 1 atmosphere meaning that any leaks or large releases of gas will fill the room from the top down. Figure 17 shows the phase diagram for Helium. Submerging objects in liquid helium allows their temperature to be reduced much further than when liquid nitrogen is used, however it is significantly more expensive to produce than liquid nitrogen. Universities with large low temperature physics groups may have a helium liquefier, although often liquid helium is purchased from an external company. If a liquefier is present, experiments will usually be connected to a recovery line so that any liquid helium which boils off from the experiment is returned to be re-liquefied.

Safety precautions are similar to those for liquid nitrogen. However, before filling a container or vessel with liquid helium, the air in the container should be removed using a rotary vacuum pump and then replaced with helium gas from a cylinder. This prevents the liquid helium solidifying any nitrogen (or small amounts of oxygen, carbon dioxide or water vapour) which may be present, leading to a blockage in the equipment.

If objects need to be cooled to 4.2 K they would usually be pre-cooled with liquid nitrogen due to the expense of liquid helium and the significantly higher latent heat of vapourisation of nitrogen. This means that only about a tenth of the mass of liquid nitrogen compared to liquid helium is needed to cool an object from room temperature to 77 K.

2.6.4 Helium-3

Helium-3 or ${}^{3}_{2}$He is a isotope of helium which has a nucleus consisting of 1 neutron (instead of the more usual 2 neutrons) along with 2 protons. At atmospheric pressure Helium-3 boils at 3.2 K (-270°C) and can't be solidified. Helium-3 is found naturally along with Helium-4, although it is approximately 10,000 time less common than Helium-4. Obtaining Helium-3 for use the in laboratory is incredibly expensive (approximately a few tens of thousands of GBP per gram in 2012) and so extreme care is taken when it is used.

The reason for its use is that it enables temperatures below around 2 K (-271°C) to be consistently and reliably achieved. If a Helium-4 bath has a large rotary vacuum pump attached to it the pressure of the vapour above the liquid can be reduced. This leads to the Helium-4 liquid cooling down to around 2 K. This is cold enough to liquify a separate jacket of Helium-3 around the sample. If a second vacuum pump is attached to the Helium-3 jacket and the pressure reduced, the

He-4 Temperature (K)	Pressure (mbar)
1.25	1.1
1.5	4.7
2.0	31
2.177	50
2.5	102
3.0	210
3.5	470
4.0	816
4.222	1013
4.5	1303
5.0	1960

Table 5: Table of temperature and vapour pressures for He-4

liquid Helium-3 will reduce in temperature to around 0.65 K. Due to the very high cost of Helium-3, great care must be taken when using it. It will have its own dedicated pump with a closed exhaust and a reservoir. When using a Helium-3 system if at all unsure what to do or what opening/closing a particular valve does always ask for help. This is much better than losing or contaminating the Helium-3.

When the pressure is reduced above a gas, the pressure can be used as a guide to the temperature of the liquid. Tables 5 and 6 give the temperature of liquid Helium-4 and Helium-3 at different pressures [5].

A more advanced use of Helium-3 and Helium-4 mixtures is in a dilution refrigerator. These can achieve temperatures of the order of a few milliKelvin but a full explanation is beyond the scope of this book.

2.6.5 Liquid Oxygen

At atmospheric pressure liquid oxygen boils at 90 K (-183°C). In its liquid form it is a translucent, pale blue liquid. Liquid oxygen is also paramagnetic, meaning that it develops an attraction to an externally applied magnetic field. Oxygen freezes creating oxygen ice at 50 K (-

He-3	
Temperature (K)	Pressure (mbar)
0.65	1.1
1.0	12
1.5	67
2.0	200
2.5	440
3.0	818
3.197	1013
3.2	1017

Table 6: Table of temperature and vapour pressures for He-3

223°C). Figure 18 shows the phase diagram for Oxygen. From the point of view of liquid oxygen being a cryogen, safety procedures are similar to those for liquid nitrogen. It is frequently thought that liquid oxygen is highly flammable and therefore very dangerous. This is misleading: liquid oxygen is an accelerant. This means that it encourages fuels to burn more quickly and intensely than they would if liquid oxygen was not present. If fuels and sources of ignition are kept away from liquid oxygen there should be no significant safety concerns in making or using small quantities.

The usual way to liquify oxygen in the laboratory is to pass oxygen gas from a compressed gas cylinder through a coil made from hollow copper pipe which is submerged in liquid nitrogen. The copper coil is a good conductor of heat and has a large surface area. Liquid oxygen is then collected in a thermos flask or a small dewar. However, the lack of a cylinder of compressed gas does not prevent liquid oxygen being made [6].

2.6.6 Argon

Argon gas is sometimes used to provide an inert atmosphere in which reactive elements or compounds can be kept for an extended period of time. At atmospheric pressure liquid argon boils at 87 K (-186°C)

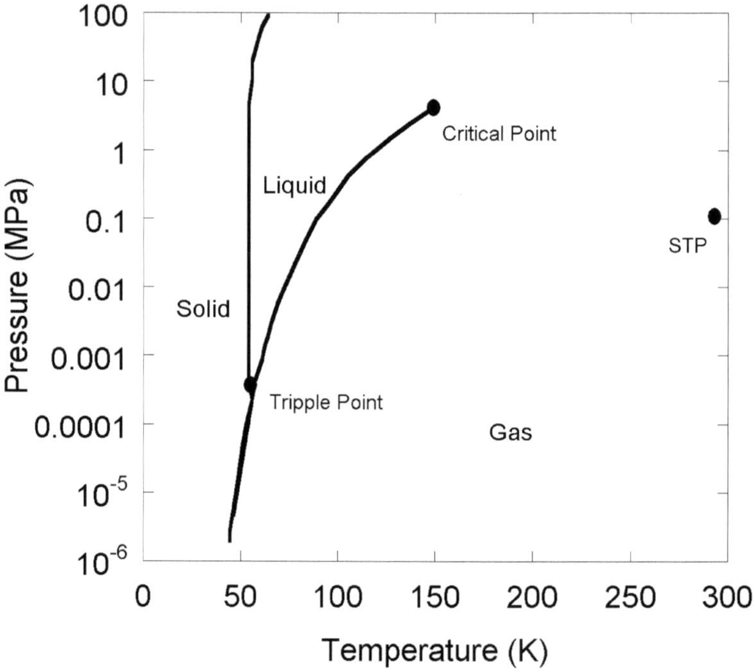

Figure 18: Phase diagram for Oxygen. Oxygen is a gas at STP (standard temperature and pressure).

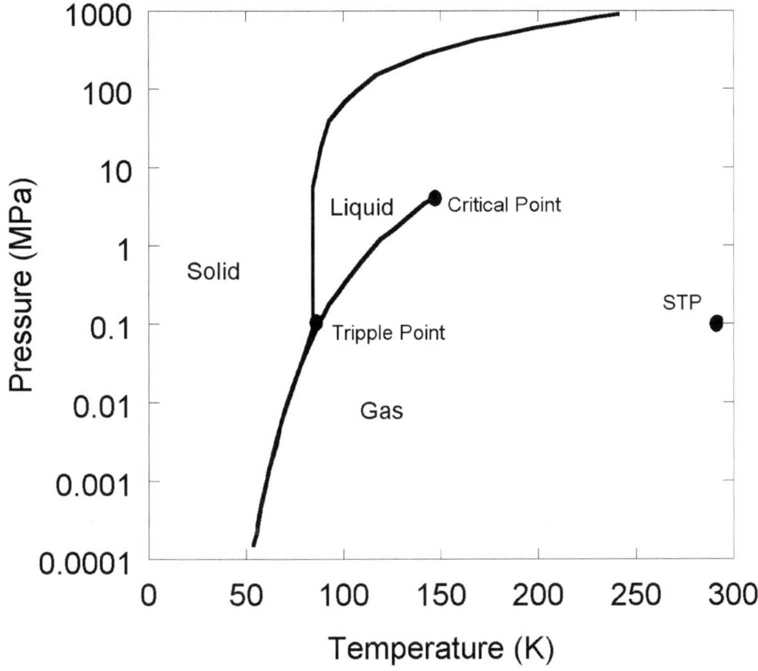

Figure 19: Phase diagram for Argon. Argon is a gas at STP (standard temperature and pressure).

meaning that it can be liquefied in the laboratory using liquid nitrogen in a similar way to the method used to produce liquid oxygen. Argon is colourless and odourless in gas, liquid and solid forms. Argon freezes creating argon ice at 84 K (-189°C).

As a gas it has a density of 1.78kg/m^3 which is greater than the density of air. This means is sinks to the bottom of a container or a room and any leaks will fill the room from the floor upwards. Figure 19 shows the phase diagram for Argon.

2.6.7 Cryogen Free Cooling

Currently cryogen free or closed cycle cryogen systems are expensive, so it is reasonably unlikely that these will met in the undergraduate physics laboratory. However, over time as prices fall and as liquid cryogens become increasingly expensive they will become more common.

Closed cycle cryogen systems aren't cryogen free. They are closed systems so that as the cryogen is boiled into a gas, the gas is collected and then recondensed using an electrically powered refrigeration system. They are sometimes classified as cryogen free as the cryogen does not need replacement.

True cryogen free systems operate on a broadly similar principle as a domestic fridge/freezer typically with helium as the working gas. Two types which are common are pulse tube refrigerators and Gifford-McMahon (GM) coolers.

2.7 Lasers

Lasers can be immensely useful in the undergraduate physics laboratory in a wide range of experiments as they provide monochromatic (single wavelength), coherent (constant phase relationship) light. Laser is an acronym for Light Amplification by Stimulated Emission of Radiation.

Although some lasers are perfectly safe, others will cause almost immediate retina damage and blindness if they are viewed even indirectly. Lasers are classified into four classes (and a few subclasses) based on their wavelength and maximum power output. Recently, a new classification system has been adopted, although lasers may still be seen labeled with the old classification system.

The new classification system is shown in Table 7.

Class 1	A class 1 laser is safe under all conditions of normal use and should not lead to eye damage even if magnified using a telescope or microscope. Devices should be fitted with a warning stating "Class 1 laser product".

Class 1M	A class 1M laser is safe under all condition of normal use except magnification with a telescope or microscope. The laser can move to a higher class if the beam is focused rather than having a large diameter or being divergent. Devices should be fitted with a warning stating "Laser radiation do not view directly with optical instruments. Class 1M laser product".
Class 2	A class 2 laser is safe because the typical human blink reflex limits the exposure. Only visible lasers (400–700 nm) can be class 2 and they are limited to 1 mW for a continuous beam. Deliberate avoidance of blinking can lead to eye damage. Many laser pointers and measuring instruments are class 2. Devices should be fitted with a warning stating "Laser radiation do not stare into beam. Class 2 laser product".
Class 2M	A class 2M laser is safe because of the blink reflex unless it is viewed through optical instruments such as telescopes or microscopes. Devices should be fitted with a warning stating "Laser radiation do not stare into beam or view directly with optical instruments. Class 2M laser product".
Class 3R	A class 3R laser is generally considered safe if care is taken to restrict viewing of the beam. Visible lasers are limited to 5 mW for a continuous beam. Devices should be fitted with a warning stating "Laser radiation avoid direct eye exposure. Class 3R laser product.".

Class 3B	A class 3B laser will cause damage to the eye if directly exposed although diffuse reflections (from matt surfaces such as walls or paper) are not harmful. Typically, visible wavelength lasers are limited to 30 mW. Protective goggles are required for the eyes where there is a possibility of direct viewing. Goggles must absorb the wavelength of light emitted by the laser and will typically be coloured. Class 3B lasers are used in CD and DVD writers, however the drive unit is Class 1 because the laser light cannot escape. Devices should be fitted with a warning stating "Laser radiation avoid exposure to beam. Class 3B laser product".
Class 4	Class 4 lasers are the most dangerous. They can cause permanent eye damage, burn skin and ignite combustible material from direct viewing and even from diffuse reflections. Great care should be taken to control the path of the beam. Many lasers in use in scientific research, medical uses and industry fall into this category. Devices should be fitted with a warning stating "Laser radiation avoid eye of skin exposure to direct or scattered radiation. Class 4 laser product".

Table 7: The new laser classifications.

The older classification system which has now been superseded may still be in use on older devices. This is shown in Table 8.

2.8 Ionising Radiation

Ionising radiation is given off by radioactive isotopes in the form of alpha and beta particles and gamma rays. There may also be positrons, X-rays and further conversion electrons emitted. Radioactive particles are present naturally in the surroundings and contribute to a background count of around 3 – 4 counts per 10 seconds if measured with a Geiger counter.

There are a range of units in which radioactivity are measured, each with slightly different meanings. The most straightforward is the Becquerel: 1 Bq corresponds to 1 nucleus disintegrating per second (note that a single nucleus may give rise to the emission of more than one radioactive particle). An alternative unit is the Curie: 1 Ci corresponds to 37 GBq. Knowing the number of disintegrations doesn't give any information about the absorbed dose a person may receive. This is measured in terms of the energy deposited per kilogram or Greys: $1\,\mathrm{Jkg}^{-1} = 1\,\mathrm{Gy}$. However the harmfulness of different types of radioactive particles varies so there is a weighting factor which is applied to the absorbed dose in Gy to give an equivalent dose in Sieverts (Sv). This weighting factor is 1 for photons, electrons and muons. It is 5 for protons >2 MeV, and neutrons <19 keV and >20 MeV. It is 10 for neutrons in the ranges 19 keV to 100 keV and 2 MeV to 20 MeV. It is 20 for neutrons in range 100keV to 2MeV and for alpha particles, fissions fragments and heavy nuclei. Finally an effective dose (also measured in Sv) is found by multiplying the equivalent dose by a tissue weighting factor accounting for the varying sensitivity of different areas of the body to radiation. A typical background radiation count over a year is $3\,\mu$Sv. In the UK,the Ionising Radiation Regulations 1999 set dose limits for the general public of 1 mSv/year, for non-classified workers of 6 mSv/year and for classified workers of 20 mSv/year.

If sources are used with thought and care there is little chance of a dose getting any where near as high as $3\,\mu$Sv which corresponds to an extra 1 year of background radiation or the dose received in a single dental X-ray. Ionising radiation can be used by anyone over 16 provided they are responsible and have had training in the safe use of the source.

Sources will commonly be sealed in a durable cup holder and kept in a lead lined box: see Figure 20. Sources should always be kept

Class I	Similar to Class 1.
Class II	Similar to Class 2.
Class IIa	Similar to Class 2, although lower power. Direct viewing for up to 1000 seconds is considered safe.
Class IIIa	Similar to Class 2M, although direct viewing for more than 2 minutes will cause retina damage.
Class IIIb	Similar to Class 3B.
Class IV	Similar to Class 4.

Table 8: The old laser classifications.

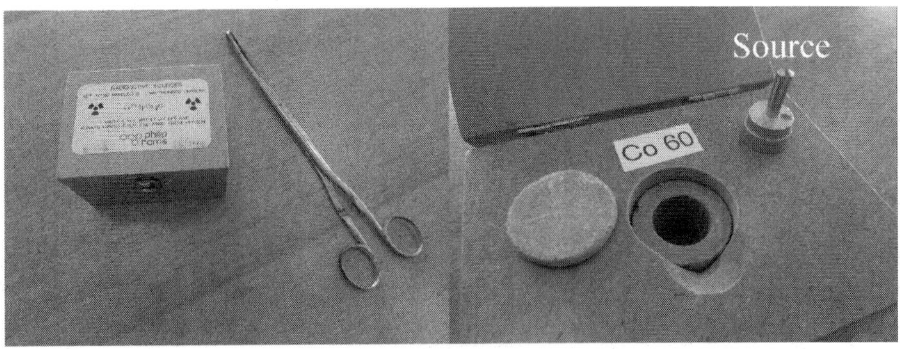

Figure 20: Left panel: a lead lined wooden storage box and forceps. Right panel: inside the box and the source inside a metal cup holder.

secure when not in immediate use. The university have a legal duty to know the exact whereabouts of every source they own: in practice this means that sources are signed out of a locked storage cabinet and then signed back in when they are no longer in use. Sources should be carried and kept as far away from the body as reasonably practical and only handled with long forceps. Be aware of leaving unused gamma sources within a few meters of an experiment as this radiation may erroneously be measured. Latex/nitrile gloves should be worn when using any sources which are not sealed in cup holders. This prevents any flakes of radioactive material becoming trapped in clothes or under finger nails.

Alpha particles have a range in air of up to a few cm. Beta particles have a significantly further range in air (around 1 m), but can be blocked by the lead lining in the source box or a few millimeters of aluminium. Gamma rays are more penetrating and will not be at all collimated. They will emit equally in all directions from the cup holder. However, gamma rays follow the inverse square law, so if the distance between the source and the user is doubled, the dose received will be quartered.

Detection of ionising radiation is usually by a Geiger-Muller tube. This is a cylinder with an axial electrode filled with low pressure gas. A potential difference of a few hundred volts is applied between the cylinder and the axial electrode. When a beta particle enters through the mica window at the front it will ionise one of the gas molecules creating a positive ion and an electron. These particles will be accelerated to opposite electrodes, when they gain sufficient kinetic energy they will be able to cause secondary ionisation events and then an avalanche effect which produces a measurable current which can be registered by a counter. Gamma particles will cause the same process, but do not need to enter the tube at the end through the mica window as they will easily penetrate the tube from any direction. Finally, alpha particles are only detected from an extremely close range and even then many

Isotope	Radiation(s)	Halflife (years)
Cobalt-60	γ (β)	5.36
Strontium-90	β	28.1
Americium-241	α (γ)	458
Plutonium-239	α	24,400
Radium-226	α, β, γ	1602
Sodium-22	β^+	2.62
Caesium-137	γ, β	30
Protactinium-234m	α, β	72 sec
Radon-220	α, β, γ	56 sec

Table 9: List of common radioactive sources. Protactinium and Radon are listed with radiation given out by the decay chain which generates them from Uranium-238 and Thorium-232 respectively.

are blocked by the mica window. A spark counter is a more appropriate way to detect alpha particles. This is a pair of wires across which a very high (electric) potential difference is applied. When an alpha source is brought near, the resulting ions created in the air causes a spark to jump between the wires.

Common sources are shown in Table 9 along with their halflife and main radiation(s).

3 Making Measurements

When using a new piece of apparatus, try to perform a preliminary experiment in which a basic measurement is made. This will increase familiarity with the equipment and confidence that it is giving a reliable result. For example, with a multimeter, the (electric) potential difference of an AA battery could be measured, expecting the outcome of around 1.5 V. Alteratively measure the resistance of a known resistor. A preliminary experiment should also allow a check to be made that the equipment is working correctly, that it has a suitable range and precision for the experiment and allow an estimate to be made of the uncertainty or error in the measurements.

This section focuses on a range of measuring devices which may be encountered, explaining how they are used and sometimes the underlying physics. Care should always be taken when reading analogue or pointer style meters and scales to avoid parallax errors. These arise when the eyes are not directly above the scale meaning the pointer may line up with an incorrect mark on the scale. Taking large numbers of decimal places from digital meters should also be avoided unless it is known the meter has been calibrated to this resolution and that the result is genuinely meaningful.

3.1 SI Units

The International System of Units is abbreviated SI from the French, Le Système International d'unités. It gives a set of seven base units in which all units can be expressed: see Table 10. For example the Newton is equivalent to the $kgms^{-2}$ and the Joule is equivalent to kgm^2s^{-2}.

3.2 SI Prefixes

Prefixes are used to show multiples and submultiples of SI units. Table 11 shows a list of values.

Unit	Symbol	Name
metre	m	length
kilogram	kg	mass
second	s	time
Ampere	A	electric current
Kelvin	K	temperature
mole	mol	amount of substance
candela	cd	luminous intensity

Table 10: The SI units

Factor	Name	Symbol	Factor	Name	Symbol
10^{24}	yotta	Y	10^{-1}	deci	d
10^{21}	zetta	Z	10^{-2}	centi	c
10^{18}	exa	E	10^{-3}	milli	m
10^{15}	peta	P	10^{-6}	micro	μ
10^{12}	tera	T	10^{-9}	nano	n
10^{9}	giga	G	10^{-12}	pico	p
10^{6}	mega	M	10^{-15}	femto	f
10^{3}	kilo	k	10^{-18}	atto	a
10^{2}	hecto	h	10^{-21}	zepto	z
10^{1}	deka	da	10^{-24}	yocto	y

Table 11: The SI prefixes

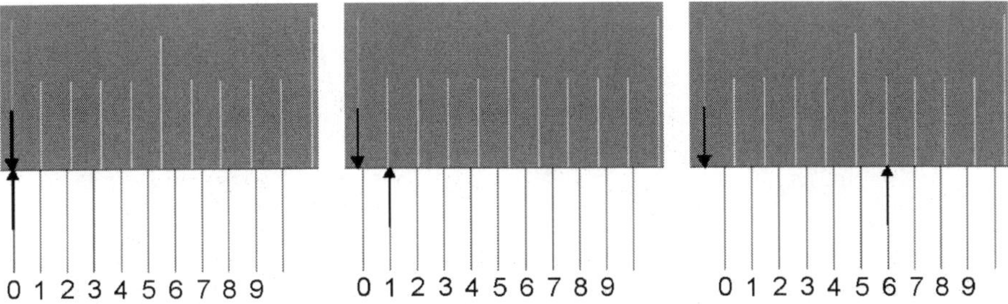

Figure 21: A Vernier Scale. The top is the main scale and the bottom in the traveling scale. The left panel shows a reading of 0.0 mm; The middle panel shows a reading of 0.1 mm; The right panel shows a reading of 0.6 mm.

3.3 Vernier Calipers

Vernier calipers (usually with a resolution of 0.01 mm) are used to measure very small distances, such as the thickness of a wire, precisely. Achieving an accurate reading with Vernier calipers is dependent on the skill of the operator. The object must always be clamped squarely between the jaws of the calipers which must be forced into contact with the object to be measured. As both the object and the calipers are slightly elastic, if too much force is applied and the object squashed the calipers will under-read. Conversely, if not enough force is used the calipers will over-read.

A Vernier scale is two adjoining traveling scales with slightly different spacings. The main scale, for example, has divisions every 1 mm. The lower scale has divisions every 0.9 mm. So 10 divisions on the lower scale take 9 mm. This means the tenth division on the lower scale lines up exactly with 9 divisions on the main scale. This can be seen in the left panel of Figure 21.

As the lower scale is moved by 0.1 mm, the first division on the lower scale will line up with the first division on the main scale. This is shown in the middle panel of Figure 21. The right panel of Figure 21 shows a reading of 0.6 mm.

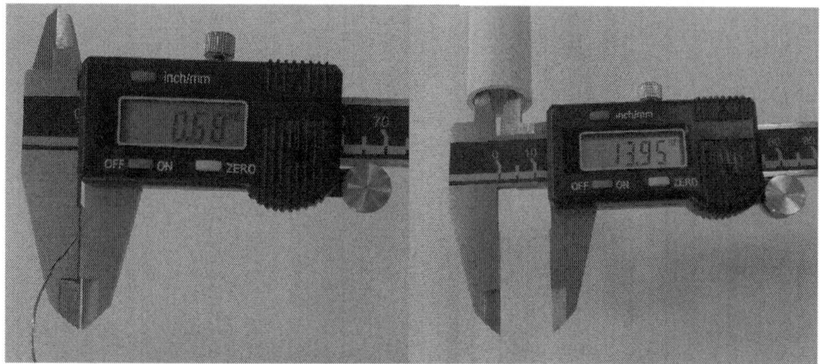

Figure 22: A digital calipers measuring; Left: the diameter of a wire and Right: the inside diameter of a pipe.

For objects larger than 1 mm, the number of millimeters on the main scale immediately to the left of the zero mark on the traveling scale must be added to the reading from the traveling scale.

3.4 Digital Calipers

Digital calipers are becoming more and more common as a replacement for Vernier calipers as they are easier and quicker to use. The jaws are closed and the reading is zeroed, usually by pressing the 'zero' button. The object to be measured can then be held in the jaws and the distance read from the display. Figure 22 shows a digital calipers being used to measure a thin wire (0.68 mm) and the inside diameter of a cylindrical pipe (13.95 mm).

3.5 Micrometer

Micrometers use a screw to convert a small linear distance moved into a large rotation which can be read from a scale. A Vernier style scale is also sometimes incorporated to increase the precision of the micrometer. The screw is usually fitted with a ratchet mechanism which allows objects to clamped with a consistent and appropriate force which reduces error due to the elastic nature of the object and the micrometer.

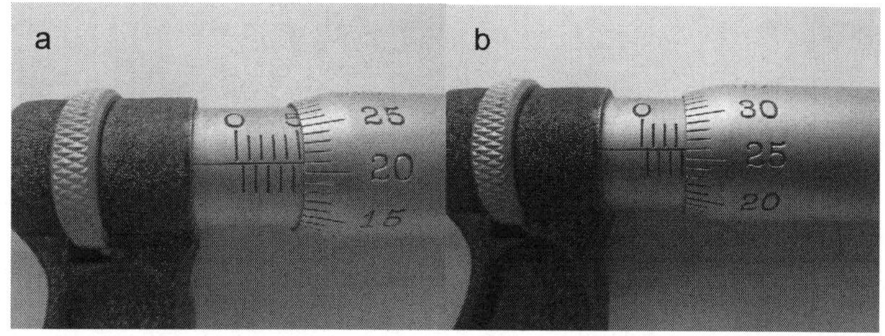

Figure 23: A micrometer measuring a: 5.21 mm and b: 3.76 mm

To make a measurement with a micrometer an object is positioned between the jaws. The ratchet screw is turned until it clicks indicating the object is correctly pressed between the jaws. A reading of the lowest half millimeter is taken from the fixed stem of the micrometer. The alignment of the rotating barrel of the micrometer with the axial line on the stem provides an extra decimal place to the reading. For example Figure 23a shows the micrometer displaying 5.21 mm and Figure 23b shows the micrometer displaying 3.76 mm.

For Figure 23a:

Stem Reading	5.00 mm
Barrel Reading	0.21 mm
Total	5.21 mm

For Figure 23b:

Stem Reading	3.50 mm
Barrel Reading	0.26 mm
Total	3.76 mm

3.6 Balances

Digital balances provide an indirect method for finding the mass of an object. Balances which measure to many mass ranges and precisions are available and it should always be ensured that the balance used is appropriate for the measurement needed. For example: it would not be appropriate to use a $0-2$ kg balance which reads to the nearest 1 g to try to measure 100 mg of metal, neither is it appropriate to check the mass of a 1 kg standard mass with a $0-500$ g balance which reads to the nearest 1mg.

Balances involve the use of one or more pressure sensors which produce an electrical output dependent upon the force acting on them. This electrical output has been calibrated by the manufacturer as corresponding to a particular force. This force is then converted to a mass by dividing by an average value for the acceleration due to gravity. According to the latest data from the NASA Grace mission the acceleration due to gravity varies around the Earth by at least $0.001\,\mathrm{ms}^{-2}$. Thus quoting masses to more than 4 significant figures is inappropriate on balances not calibrated for use in a specific location, although this may not be true for mass differences.

The balance should be positioned directly on a flat surface: ensure there are no wires or other objects under the balance. When making measurements on a table, desk or worksurface ensure that no one touches, leans or sits on the surface as this can sometimes significantly alter the reading on the scale. It is useful to perform a quick experiment to see the significance of this. Some balances have a cover to prevent draughts (possibly from people walking past or convection currents from heating or air conditioning) affecting the results: this should be placed down and closed when any readings are taken.

To take a measurement first ensure that the scale pan is clean and free from loose material and the balance reads zero. If not the 'Tare' or 'Zero' button should be pressed. If a container is used often to hold powder which is being measured out, this can be placed on the balance before it is zeroed. Finally add the object to be measured to the scale pan and take a reading.

3.7 Ammeters

Ammeters measure the current flowing in a circuit. Current is the rate of flow of charge. They are placed in series with the component through which the current is to be measured. To cause the minimum (or no) change to the behaviour of the rest of the circuit an ideal ammeter will allow current to flow through with no resistance. In practice ammeters have a very low resistance, which is assumed to be small compared to the circuit being measured. To picture this more clearly consider the equation for the total resistance of resistors in series:

$$R_T = R + R_{ammeter} \tag{3}$$

It is required that R_T is as close as possible to the resistance, R of the component. This can be achieved by making $R_{ammeter}$ as small as possible.

3.8 Voltmeters

Voltmeters measure the (electric) potential difference (sometimes called the voltage drop) across a component or between two points in a circuit: this is the difference in energy per coulomb of charge. They are placed in parallel with the component across which the (electric) potential difference is to be measured. To cause the minimum (or no) change to the behaviour of the rest of the circuit an ideal voltmeter will have an infinite resistance and allow no current to flow through it. In practice voltmeters have a very high resistance, which is assumed to be large compared to the circuit being measured. To picture this more clearly consider the equation for the total resistance of resistors in parallel:

$$\frac{1}{R_T} = \frac{1}{R} + \frac{1}{R_{voltmeter}} \tag{4}$$

It is required that R_T is as close as possible to the resistance, R of the component. This can be achieved by making $R_{voltmeter}$ as large as

Colour Colour	1st Band	2nd Band	3rd Band Multiplier	gap	4th Band Tolerance
Black	0	0	$\times 1$		
Brown	1	1	$\times 10$		$\pm 1\%$
Red	2	2	$\times 100$		$\pm 2\%$
Orange	3	3	$\times 1{,}000$		
Yellow	4	4	$\times 10{,}000$		
Green	5	5	$\times 100{,}000$		$\pm 0.5\%$
Blue	6	6	$\times 1{,}000{,}000$		$\pm 0.25\%$
Purple	7	7	$\times 10{,}000{,}000$		$\pm 0.1\%$
Grey	8	8			
White	9	9			
Silver					$\pm 10\%$
Gold					$\pm 5\%$

Table 12: Resistor colour codes

possible.

3.9 Resistor Colour Codes

Resistors typically have 4 coloured lines indicating the size of the resistance. Table 12 indicates how to find the resistance.

3.10 Multimeters

Frequently ammeters and voltmeters used in the lab will actually be part of a multimeter device. Multimeters are versatile devices capable of making measurements of variables such as current, (electric) potential difference in both AC and DC circuits and resistance. Figure 24 shows two typical multimeters. The round dial in the centre selects which variable the meter measures as well as the maximum reading measurable in that range. The Ω symbol indicates a resistance measurement, the A symbol indicates an ammeter and the V symbol indicates a voltmeter. The symbol of a solid line above a dashed line indicates DC and the sin wave symbol indicates AC. The setting $20\,\mu$A indicates that the maximum reading measurable will be $20\,\mu$A and the

Figure 24: Two different multimeters.

reading given will be in units of μA. The setting 200 kΩ indicates that the maximum reading measurable will be 200 kΩ and the reading will be given in units of kΩ.

Two wires must be connected to the front of the multimeter: one to the black common point and the other to a red point corresponding with the variable being measured.

To measure a resistance using only two connections the multimeter generates a known constant current which it passes through the component or circuit connected to the meter. It then acts as a voltmeter and measures the potential difference across the component. The simple calculation of $R = V/I$ gives the resistance.

3.11 Further Ideas on Measurement of Resistance

The usual method of attaching two wires to a component/sample of material to measure its resistance involves passing current and measuring the (electric) potential difference through the same wires. This has the disadvantage of measuring the sum of the component/sample

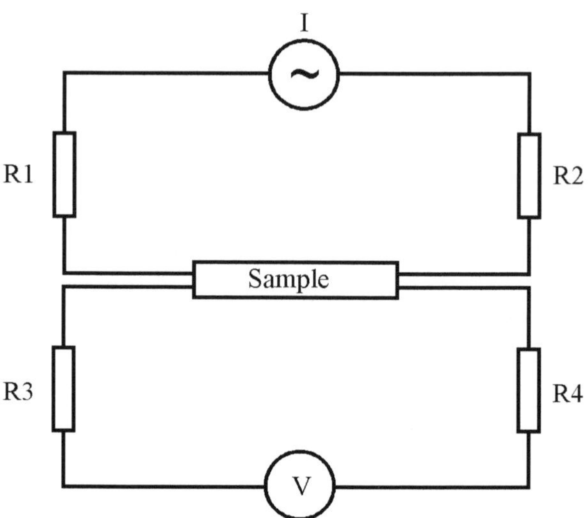

Figure 25: Simple diagram of a four-wire method of measuring the sample resistance. R1, R2, R3 and R4 are resistances which represent the contact and lead resistances.

resistance, contact resistance and the resistance of the measurement wires. In some cases where the component/sample resistance is much higher than the lead resistance this is acceptable as the additional resistance only forms a very small part of the measurement. In the case of samples with resistance $O(\mathrm{m}\Omega)$ an alternative method must be found especially if the measurement wires have a high resistance (perhaps due to their narrow diameter to reduce the heat load on a cryogenic experiment).

Resistance can be measured by using the four wire method (Figure 25) where four wires are separately attached to the sample. Two are used to pass current through the sample and a further two are used to measure the (electric) potential difference across the sample. The source current flows through R1, the sample and R2 and because no current flows through R3 and R4 the (electric) potential difference measured by the voltmeter is just that across the sample.

Figure 26 shows a contact configuration which can be used to mea-

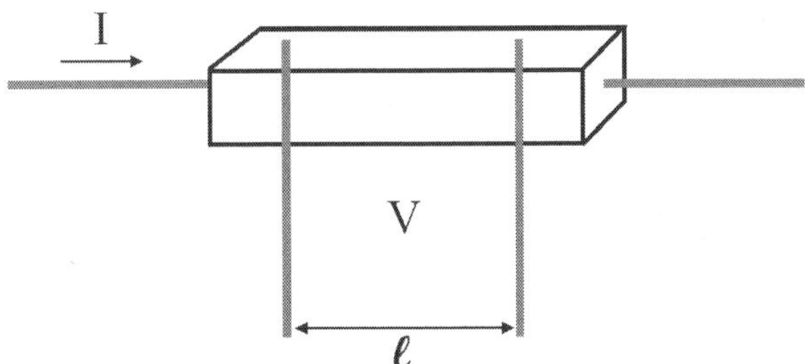

Figure 26: Simple four-wire mounting configuration of a sample to measure resistance along in the direction of the current flow.

sure the resistance of a sample of conductive material. The current is passed through the sample via the two surfaces at each end and the (electric) potential difference is measured over a distance ℓ between the other two contacts. The voltage contacts should be far away from the current contacts to ensure the current flow is uniform and parallel between the voltage contacts. The current contacts should cover the whole of the end surfaces to ensure a uniform flow.

3.12 Constant Current Sources

When making electrical measurements in which the resistance of the sample might change, a constant current source is needed. This is not trivial to construct as a cell, battery or normal power supply all have internal resistance. Recalling that:

$$I = \frac{\epsilon}{R + r} \tag{5}$$

where I is the current, ϵ is the EMF of the power source, R is the external resistance and r is the internal resistance of the power source. If R changes, but ϵ and r remain constant then the current in the circuit changes. R and I have a non linear relationship which makes calculation of R given V and I very difficult.

An ideal current source should output the same current regardless of the resistance of the sample or the leads. A number of different methods of generating constant current can be used.

The most straight forward method is to use a variable resistor R_{series} in series with the sample. An AC or DC supply can be connected to this. Alternatively, if a lock-in amplifier is used they have an internal oscillator which outputs an AC emf in the range $0-5$ V. Typically resistors between $R_{\text{series}} = 1\,\text{k}\Omega$ and $10\,\text{k}\Omega$ can be connected to the lock-in amplifier front panel oscillator output and to the sample as shown on the left in Figure 27. It is easily seen that the current flowing through the sample is given by:

$$I = \frac{V}{R} = \frac{V}{R_{\text{series}} + R_{\text{sample}}} \tag{6}$$

$R_{\text{series}} \gg R_{\text{sample}}$ otherwise as the sample resistance changes for any reason (perhaps due to a change of temperature) the current going through the sample will also change leading to ambiguous results. The value of R_{series} used is limited by the fact that most lock-in amplifiers can only output 5 V; when a larger R is used, the current becomes very small. Currents need to be sufficiently large to produce a high quality signal above any background noise as well as sufficiently low so that no self heating is detectable in measurements.

An alternative method is to use the (electric) potential difference controlled current source circuit as shown on the right in Figure 27. For an op-amp with the non-inverting(+) input connected to ground, the (electric) potential difference between the inverting(-) input and ground is also zero. An ideal op-amp has an infinite input impedance so draws no current from its input circuit implying the sum of the currents (Kirchoff's First Law) at point A should be zero:

$$\frac{V_{\text{in}}}{R_i} = -\frac{V_B}{R_f} \tag{7}$$

Where V_{in} is the (electric) potential difference from the lock-in oscillator output; R_i and R_f are the values of the resistors and V_B is the

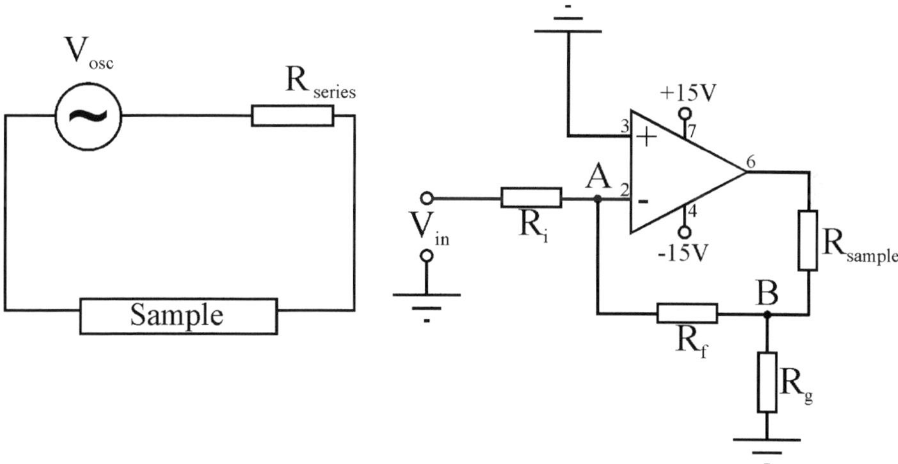

Figure 27: Left panel shows a circuit layout of a resistor (typically $R_{\text{series}} = 1\,\text{k}\Omega$ to $10\,\text{k}\Omega$) used as a constant current source. It is connected in series with the sample and the oscillator output on a lock-in amplifier. Right panel shows a circuit diagram of a (electric) potential difference controlled constant current source. The op-amp used is an Linear Technology LTC1150CN8#PBF (RS order number 5455629).

(electric) potential difference between point B and ground. Summing the currents going through the sample gives:

$$I_{\text{sample}} = \frac{V_B}{R_g} + \frac{V_B}{R_f} \tag{8}$$

Combining equations 7 and 8 gives I_{sample} which is independent of the sample resistance and is proportional to V_{in}.

$$I_{\text{sample}} = -\frac{R_f + R_g}{R_i R_g} V_{\text{in}} \tag{9}$$

The lock-in oscillator output is connected to V_{in} and the sample current contacts are those attached to the op-amp output. The current in the sample is given by equation 9.

A final method of generating a constant current is to use a dedicated off the shelf constant current source such as a Keithley 6221 AC and DC current source. This has two current output leads which are connected to the sample.

3.13 Lock-in Amplifers

At a very simplistic level this can be thought of as a voltmeter. It accurately measures very small AC signals by using phase-sensitive detection to extract the signal from an often huge background that would otherwise obscure the signal.

The sample is excited by an AC current at fixed frequency, ω_r, and the lock-in detects the response of the sample at the same frequency along with any background noise, $\sum_f V_{f,\text{sig}} \sin(\omega_f t + \theta_{f,\text{sig}})$. It uses a phase sensitive multiplier to multiply this by a self generated reference signal at the same frequency as the current $V_L \sin(\omega_r t + \theta_{\text{ref}})$.

$$V_{\text{psdoutput}} = \sum_f V_{f,\text{sig}} V_L \sin(\omega_f t + \theta_{f,\text{sig}}) \sin(\omega_r t + \theta_{\text{ref}}) \tag{10}$$

$$V_{\text{psdoutput}} = \sum_f \frac{1}{2} V_{\text{sig}} V_L \cos([\omega_r - \omega_f]t + \theta_{\text{sig}} - \theta_{\text{ref}})$$
$$- \frac{1}{2} V_{\text{sig}} V_L \cos([\omega_r + \omega_f]t + \theta_{\text{sig}} - \theta_{\text{ref}}) \quad (11)$$

For each component frequency, ω_f, that is detected this output is two AC signals, one at the difference frequency $\omega_r - \omega_f$ and the other at the sum frequency $\omega_r + \omega_f$, which are then passed through a low pass filter. In the general case for ω_r and ω_f there is no signal from the low pass filter as both signals are AC and are filtered out. If $\omega_r = \omega_f$ then part of the signal is no longer AC, but DC and proportional to the signal amplitude $V_{\text{lpfoutput}} = \frac{1}{2} V_{\text{sig}} V_L \cos(\theta_{\text{sig}} - \theta_{\text{ref}})$. This gives the X output of the lock-in. The Y output is obtained by using a second phase sensitive detector to multiply the detected signal by the reference frequency shifted by 90° which gives an output $V_{\text{lpfoutput}} = \frac{1}{2} V_{\text{sig}} V_L \sin(\theta_{\text{sig}} - \theta_{\text{ref}})$.

Changing the offset of the reference frequency, θ_{ref}, alters 'the phase' $(\theta_{\text{sig}} - \theta_{\text{ref}})$ of the measurement. For a pure resistor in an AC circuit the voltage response is exactly in phase with the current so the phase should be set to zero $(\theta_{\text{sig}} - \theta_{\text{ref}} = 0)$ meaning that the X ('in phase') output of the lock-in gives the measured signal $(\cos(0) = 1)$ and the Y ('out of phase') output is zero $(\sin(0) = 0)$. For an inductor the current lags the voltage by 90° and for a capacitor the current lead the voltage by 90° thus both these will show up on the Y ('out of phase') output.

The frequency of the excitation current, ω_r, can be altered on the lock-in. Typically it would be in the region 50 – 100 Hz: but taking care to avoid the mains electricity frequency (50 Hz in UK and 60 Hz in USA) and its harmonics (factors or multiples).

Figure 28 shows the front panel of a Stanford Research Systems (SRS) 830 lock-in amplifier. Usually BNC cables would be connected to the 'A' and 'B' inputs in the bottom left of the front panel and to the 'SIN OUT' or 'REF IN' BNC connectors at the bottom right of the front panel.

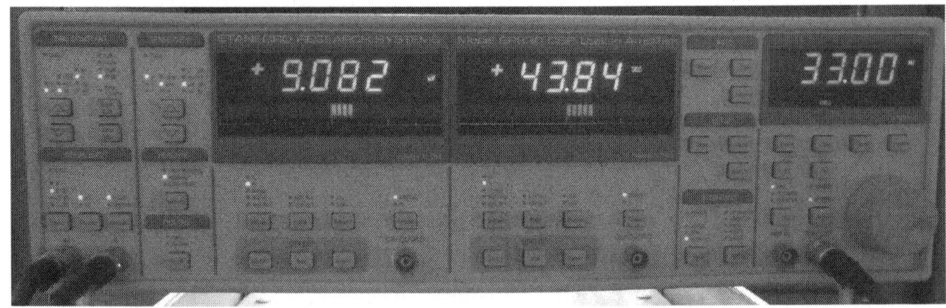

Figure 28: The front panel of a Stanford Research Systems (SRS) 830 lock-in amplifier.

First, a frequency must be selected for the lockin amplifier to use. The right hand display shows the frequency when the 'Freq' button is pressed. It can be adjusted by rotating the dial on the right of the front panel. Often the internal frequency generator is used which outputs a sine wave via the 'SIN OUT' connector. This is then used within the circuit – for example see Section 3.12 on constant current sources. If the internal frequency generator is used, the amplitude can be displayed by pressing the 'Amp' button and can be changed by rotating the dial.

The frequency can also be selected by supplying a sine wave or squarewave to the 'REF IN' input. The 'Source' button can be pressed to switch between the 'REF IN' signal and the internal frequency. When the lock-in amplifier has not identified a signal to lock on to the 'UNLOCK' light is illuminated.

The left display (channel 1) would usually be set to show the 'X' input which is the in phase component of the signal. The right display (channel 2) would then be set to show the 'Y' input which is the out of phase component of the signal. The parameter displayed can be changed by pressing the 'Display' button.

The sensitivity panel displays the maximum (electric) potential difference which can be measured in the present range. The range can be changed by using the arrow keys to cycle through the available sensitivities/ranges. Underneath the X and Y display there is a red 'bar graph'. This gives a visual representation of size of the signal in proportion to the present range. The range does not change automatically and so needs to be adjusted as the signal amplitude changes.

The time constant setting gives the characteristic time period over which the outputs are averaged. A long time constant can be used to smooth out very noisy signals, but this will increase the response time.

To measure the true signal from a sample the phase should be set correctly. The right hand display shows the phase when the 'Phase' button is pressed. For a resistance measurement, as a first approximation, assume the sample is purely resistive and the experimental equipment is sufficiently well designed that it doesn't cause large capacitive or inductive signals. The phase of the measurement can be set using the 'Phase' button on the 'Auto' panel of the lock-in. This sets the phase such that the Y output is zero. In most cases this is a perfectly adequate assumption to make. In some cases such as when the sample contacts are not of optimal quality the phase of the measurement needs to be set to account for the phase shift of the true resistive component of the signal. Firstly, the normal measurement circuit shown in the left panel of Figure 29 is modified to include a series resistor, approximately of the same resistance as the sample giving the circuit in the right panel of Figure 29. The (electric) potential difference across this resistor is then measured with the lock-in. In the ideal case this is a pure resistor so the auto-phase button is pressed to set the phase such that the out-of-phase component, Y, is zero, the phase is then kept at this value when the circuit is then returned to the original configuration for the experiment.

3.14 Further Ideas on Reducing Noise in Electrical Measurements

There are a number of possible sources of noise/error which might affect electrical measurements such as external electromagnetic fields, ground loops, capacitive coupling, self-inductive effects and inductive

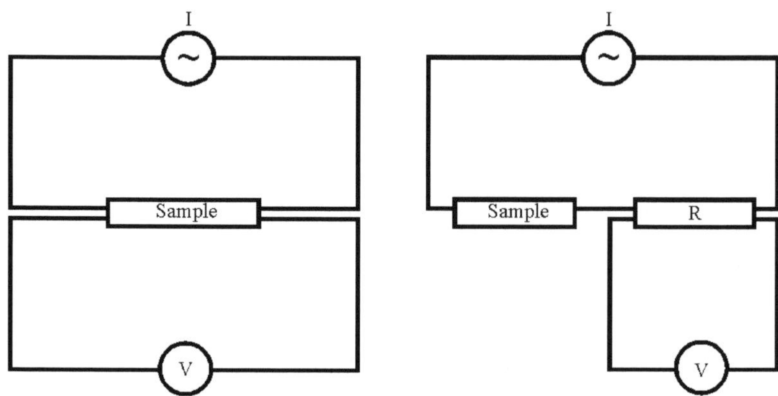

Figure 29: Left panel shows the circuit setup for a normal 4-wire measurement configuration. Right panel shows the circuit setup when setting the phase. Current is passed though the sample and a series resistor, R. The (electric) potential difference drop is measured across only the resistor.

cross-talk between wires.

The total impedance of a circuit is given by the sum

$$\begin{aligned} Z &= \sqrt{X^2 + Y^2} && (12) \\ &= \sqrt{R^2 + X_C^2 + X_L^2} && (13) \\ &= \sqrt{R^2 + \left(\frac{1}{\omega_r C}\right)^2 + (\omega_r L)^2} && (14) \end{aligned}$$

where R is the resistance, X_C is the reactance due to the capacitive voltage component with a phase which lags the current by $\pi/2$ and X_L is the reactance due to the inductive voltage component which leads the current by $\pi/2$. This means that the Y output on the lock-in amplifier contains all the (electric) potential differences of capacitive and inductive origin and is dependent on the frequency of the excitation current ω_r and the X output contains just the resistive component and is independent of frequency. Clearly the resistive, capacitive and inductive components can also be temperature dependent.

The effect of external fields can be reduced by shielding as much of

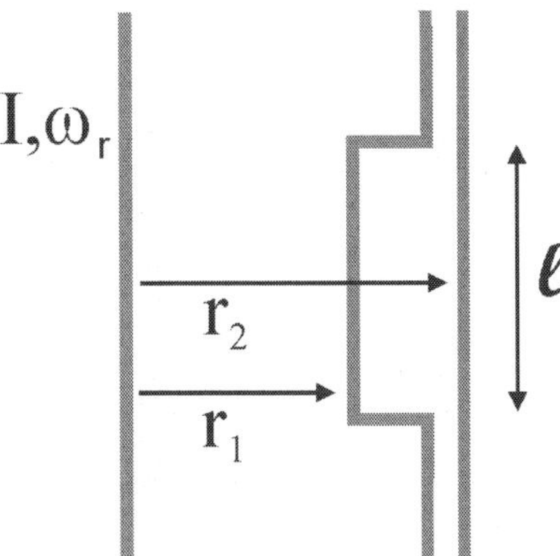

Figure 30: Inductive cross-talk between wires in two closed circuits carrying AC signals.

the wiring as possible. This takes the form of a grounded fine woven metal braid and/or metal foil surrounding a set of wires and is built into cables used to connect experimental equipment. Static electromagnetic waves are unable to penetrate a volume of space entirely surrounded by a conducting shield and oscillating electromagnetic waves undergo an exponential decay through the shield [7].

All cable shielding and grounds of equipment should be connected together to a single grounding point. Alternatively the cable shielding is sometimes broken in a certain place to isolate two true ground points from each other. If great care is not taken over this then there may be more than one grounding point in the circuit. An (electric) potential difference can arise between these points which can cause large noise currents to flow and adversely affect the quality of the measured signal significantly.

Inductive cross-talk is noise resulting from the mutual inductance of two or more closed circuits with paths located near each other i.e. the changing magnetic field created by an alternating current in one circuit creates spurious signals in a neighboring circuit according to Faraday's law $V = d\Phi/dt$. Figure 30 shows an arrangement of wires that are subject to cross-talk. In the left wire a current I with frequency ω_r

flows. The current creates a magnetic field at distance r from the wire at time t given by

$$B(r,t) = \frac{\mu_0 I \sin \omega_r t}{2\pi r} \tag{15}$$

The other two wires are part of a neighboring circuit and situated parallel to the first wire a distance of r_1 and r_2 away. The separation of these two wires $(r_2 - r_1)$ continues for length ℓ. Ampere's law can be used to show the magnetic flux induced through this area of the circuit is

$$\Phi = \oint_S \mathbf{B}(r,t) \cdot d\mathbf{A} \tag{16}$$

$$= \int_{r_1}^{r_2} \int_0^l \frac{\mu_0 I \sin \omega_r t}{2\pi r} dz dr \tag{17}$$

$$= \frac{l \mu_0 I \sin \omega_r t}{2\pi} \ln \frac{r_2}{r_1} \tag{18}$$

Using Faraday's law shows the (electric) potential difference this causes is proportional to ω_r and $\ln r_2/r_1$

$$V = -\frac{d\Phi}{dt} \tag{19}$$

$$= -\frac{l \mu_0 I \omega_r \cos \omega_r t}{2\pi} \ln \frac{r_2}{r_1} \tag{20}$$

To minimise emfs induced in this way the frequency of the AC current is kept low and the loop area between wires is kept to a minimum by twisting both wires of a circuit together creating twisted pairs. Another important reason to keep loop areas to a minimum arises when measurements in a magnetic field are considered. Any current carrying loop of wire in a magnetic field will be subject to a torque $(T = IAB)$ proportional to the current I, the loop area A and the magnetic field B. As the current is AC, this will cause the loop to vibrate and induce an emf at the same frequency, ω_r as the current. The best way to minimise this is to securely stick/tie down all wires especially those which are very thin and light.

Lock-in amplifiers use a differential amplifier whose ideal output is proportional to the (electric) potential difference between its inputs i.e. the (electric) potential difference across the sample $V = Gain \times (V_+ - V_-)$. A real amplifier also has an output component proportional to the actual (electric) potential differences $V = Gain \times \left((V_+ - V_-) + 10^{-6}\frac{(V_+ + V_-)}{2}\right)$. This additional component is the common-mode (electric) potential difference. When the (electric) potential difference measured by the lock-in is very small this common-mode leak (electric) potential difference can adversely affect the results. The best way to minimise the common-mode (electric) potential difference is to balance the resistance of each wire in the circuit with its pair.

3.15 Signal Generator

A signal generator produces a time varying emf output. The frequency is selected by choosing a range using the range dial and then setting the frequency dial. The amplitude of the wave can be increased or decreased using the amplitude/volume control. The format of the wave can be chosen from a sinusoid, a square wave, a triangular/saw tooth wave or the TTL output. Figure 31 shows an example of a signal generator.

3.15.1 TTL

The TTL (Transistor-Transistor Logic) output provides a standard signal between 0 V and 5 V. A TTL output signal is defined as "low" when between 0 V and 0.4 V with respect to the ground terminal, and "high" when between 2.4 V and 5 V. A TTL input signal is defined as "low" when between 0 V and 0.8 V with respect to the ground terminal, and "high" when between 2 V and 5 V. These signal are used to pass timing or triggering signals between different pieces of equipment such as signal generators and oscilloscopes or current sources and lock-in amplifiers.

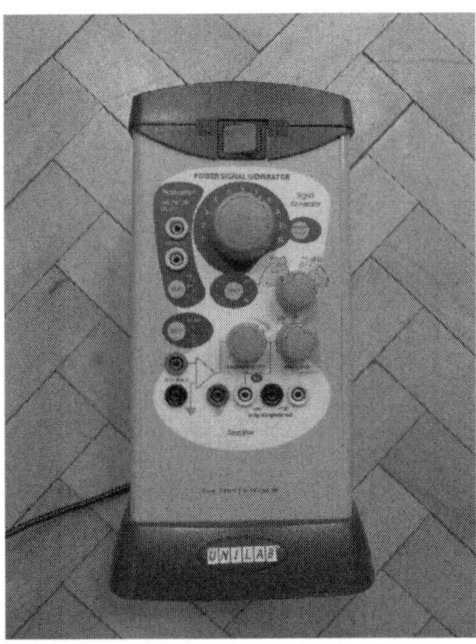

Figure 31: A signal generator.

3.16 Oscilloscopes

Oscilloscopes are enormously useful pieces of equipment in the laboratory provided their operation is understood. Figure 32 shows the front panel of an analogue or CRT oscilloscope and Figure 33 shows the front panel of a digital oscilloscope. The screen on an oscilloscope can be loosely thought of as a graph. It is usually marked with a graticule scale of large squares with length 1cm, as well as smaller sub divisions. The horizontal axis is usually the time and the vertical axis is the electric potential difference.

There are two inputs for BNC cables: Channel 1 (CH1) and Channel 2 (CH2). BNC cables provide a firm connection which does not break when pulled. The metal cap on the wire must be pushed towards the instrument and twisted anticlockwise (clockwise) to disconnect (connect) the cable. Cables are usually coaxial which means they have a core down which the signal travels. This is surrounded by an insulating wrapper and then anther conducting wrapper which may be foil or braid and acts as an electrical shield. Finally this is surrounded

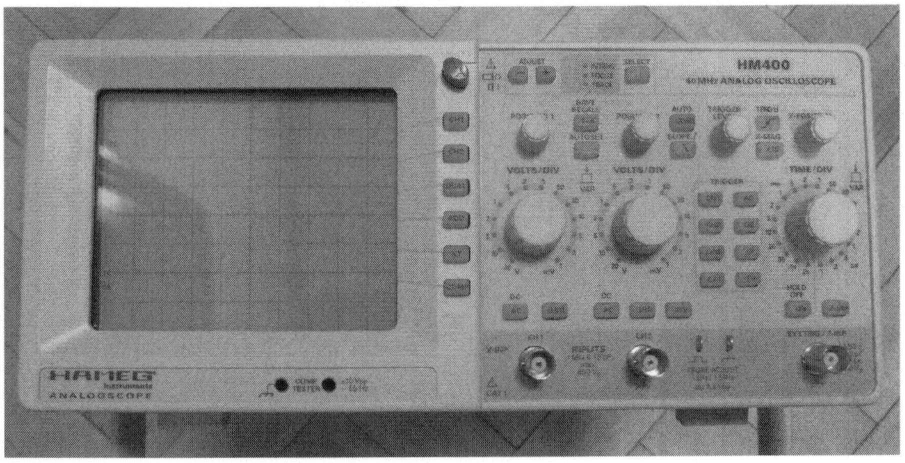

Figure 32: An CRT oscilloscope.

by another insulating sheet.

The following list gives a brief description about the function and use of common oscilloscope buttons.

- **'CH1 Volts/Div' and 'CH2 Volts/Div'** also sometimes called the 'Vertical Gain'. There is one dial per channel, which controls the vertical amplitude of the waveform. The settings are usually given in terms of Volts per cm. So decreasing the Volts per cm, increases the gain and the amplitude of the waveform on the screen. The vertical gain is normally set so that the waveform fills the screen from top to bottom without going off the screen. There are three common sources of confusion in checking amplitude measurements are as expected.
  - AC (electric) potential differences are usually discussed in terms of RMS (root mean square) values. Oscilloscopes display peak values which are $\sqrt{2} = 1.41$ times larger than RMS values.
  - Sometimes there are calibration controls in the center of the Volts/Div and timebase dials. These should be in a normal

position indicated by a 'locking' felt at the dial is rotated.

 - Sometimes special probes are used on an oscilloscope which scale the signal by a factor of 10 or 100. This will be printed on the probes attached to the CH1 and CH2 inputs.
 - Some oscilloscopes have a button to magnify the x and/or y axis by a factor of 10 independently of the volts/div or timebase controls.

- **Vertical Position** This controls the vertical position of the waveform on the screen. When no signal is present the trace on the screen is usually adjusted so that it sits on a convenient graticule marking on the screen so that amplitudes above and below zero can be easily measured. When taking amplitude measurements it is sometimes convenient to adjust the vertical position to line up the peak and trough of the waveform with the graticule lines.

- **Timebase** The timebase or SEC/DIV dial controls the speed at which the screen is scanned. The settings are usually given in terms of the number of seconds per cm. The frequency of the waveform can be calculated from this by taking one divided by the time period for one wavelength. For example if the time base setting is 0.01 seconds/cm and one complete wavelength takes 5 cm, the period is 0.05 seconds and the frequency is $1/0.05 = 20$ Hz. Similar to the vertical gain, the timebase is set so that the waveform fills the screen with a small number of periods from right to left.

- **Horizontal Position** This is similar to the vertical position control, but changes the horizontal position. This can be adjusted when taking time measurements so that the waveform lines up well with the graticule lines.

- **Autoset** Some modern oscilloscopes have an autoset button. When this is pressed, the CH1 gain, CH2 gain and the timebase are automatically set to appropriate values. This is an especially useful feature to have as it saves a great to deal of wasted time when oscilloscope settings are unfamiliar or the frequency/magnitude of the signal is unknown.

- **Trigger** The trigger setting controls the point at which the waveform begins to appear on the left hand side of the screen. Each time the waveform reaches the right hand side of the screen the oscilloscope will pause and wait until a certain condition is met before starting again at the left hand side of the screen. For periodic functions this provides a synchronised waveform which does not drift across the screen and it enables randomly occurring pulses to be displayed easily. The trigger level control sets the (electric) potential difference which the input has to reach before the display of the waveform is restarted. Other trigger related buttons can control when the trigger is based on channel 1 or channel 2 or whether it is based upon the rising or falling edge of the waveform. The trigger can also be set to external. This triggers the oscilloscope based upon a pulse from an external source such as the TTL output from a signal generator or lock-in amplifier.

- **Mode** There are usually a series of buttons or a dial which sets what is displayed on the screen (in Figure 32 they are down the right hand side of the screen). This allows a choice between displaying only channel 1 (CH1), only channel 2 (CH2), both channel 1 and 2 at the same time (DUAL), the sum on channel 1 and channel 2 (ADD). The XY mode displays the CH1 input as the x coordinate and the CH2 input as the Y coordinate. If two different AC signals (at frequencies f_1 and f_2) are connected to CH1 and CH2 when the oscilloscope is in XY mode, interesting patterns called Lissajous figures appear when $f_2 = nf_1$.

- **Coupling** The GND/DC/AC buttons for CH1 and CH2 alter

the coupling between the signal input and the input to the amplifier. The DC setting means that the input is connected directly to the amplifier. The AC setting means that a capacitor is connected between the input and the amplifier so that DC (electric) potential differences are blocked. The GND setting means that the amplifier input is connected to the ground (0 V) which allows a check of the position of 0 V on the display. The DC position should be appropriate in most cases.

- **Measurements** More modern, digital oscilloscopes have horizontal and vertical cursors or bars which can be displayed on the screen. Usually the button is labeled 'Measure'. The cursors can be moved and the screen displays the (electric) potential difference between the horizonal cursors and the time period between the vertical cursors. It is also possible to get the oscilloscope to calculate and display the period, frequency and amplitude of the signals.

To begin using an oscilloscope first find and operate the power button on the front panel. It is then usual for a power LED to turn on and perhaps a number of other LEDs. If nothing happens, check for an additional power switch on the back of the oscilloscope and that the mains lead is properly connected and that the mains is turned on.

Wait for the beam to be displayed. An analogue oscilloscope may take a moment or two to turn on and a digital one will take longer as the software starts up. If the trace is not displayed on the screen then check the trigger level dial is set to the center and the horizontal and vertical position controls are set to the center.

Next connect the signal via the BNC connector on the front panel to at least one channel. The coupling should be set to DC in most cases and the mode set to either CH1, CH2 or DUAL depending on where the input signal(s) are connected. Set the vertical gain and timebase dials to an appropriate setting so that the waveform occupies a large part of

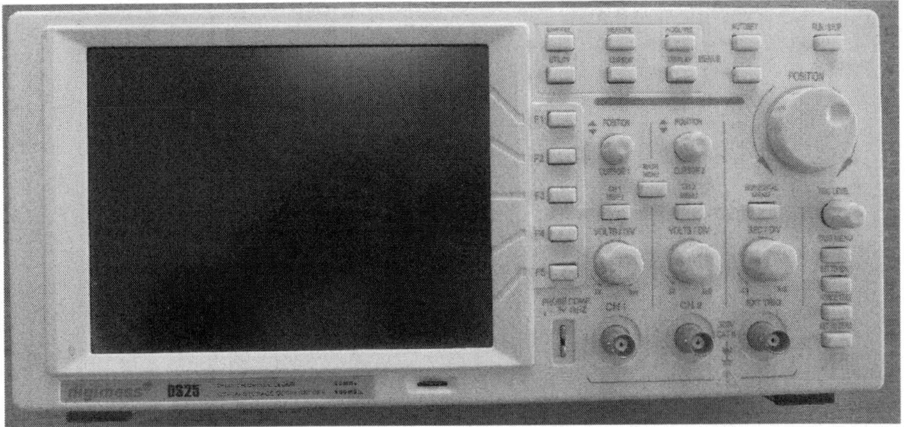

Figure 33: A digital oscilloscope.

the screen. If no signal appears, try pressing the autoset button if the oscilloscope has one otherwise the BNC cables and any other cables should be checked. If there appears to be poor connection, then the cables should be swapped for a different set. If this fails the oscilloscope could be connected to a signal generator which is known to be working.

Finally the trigger should be set to one of the channels with a signal and the level control can be adjusted along with the rising or falling edge selection so that the best waveform is displayed.

3.17 Light Gates

Light gates are very useful for measuring time, speed/velocity and acceleration of moving objects. To simplify and improve the flow of the language, in this section it will be assumed the object is a toy car. Of course, in reality it may be any other suitable moving object. It is vital to point out that light gates are timers. The only measurement they make is the time for which a light beam (usually infra red) is broken. Further processing and thoughtful setup is required to find the velocity and acceleration using either one or two light gates. Often, the light gate control box or computer will do much of the processing, but will initially need to be programmed with a number of distance measurements.

3.17.1 Velocity with One Light Gate

Usually a small, vertical piece of card of known length would be fixed to the top of the car. The velocity of the car at a given point is found by placing the light gate at the point and dividing the length of the card by the time taken for the card to travel through the light gate. If a piece of card is not used, it must be clear the exact length of the car which is breaking the light beam.

3.17.2 Velocity with Two Light Gates

In this case a piece of card is not necessary and the method may be more suitable where card can not easily be used. The two light gates are setup a known short distance apart. The light gates are setup to measure the time between the gates, so that the timer is started as the car travels through gate A and is stopped when the car travels through gate B. The velocity is found by dividing the distance between the light gates by the measured time. This gives an average velocity over the distance between the two light gates and is similar to how average speed checks on motorways in the UK work.

3.17.3 Acceleration with Two Light Gates

Two velocity measurements taken a known time apart are needed. Each of these can be taken in a similar way to the measurement of the velocity with one light gate. The average acceleration of the car between the two light gates is found by diving the difference in the two velocity measurements by the duration of time between them.

3.17.4 Acceleration with One Light Gate

To measure the acceleration with just one light gate, two velocity measurements are still needed. These can be taken using a 'U' shaped card. The first velocity is measured using the first card to break the beam and the second velocity is measured using the second card to break the beam. Here it is important to measure the width of each of the vertical sections of the 'U'. The acceleration is calculated in a similar way by diving the difference in the two velocities by the duration of time between them.

3.18 PASCO Science Workshop Interface

PASCO Scientific produce a range of interface boxes to connect a measurement devices to a computer for data logging. One example of this is the Science Workshop 500 Interface see: Figure 34. This box connects to the mains via a transformer and to the computer via a USB to serial converter. A large selection of sensors are available such as: photogate, ultra sound motion sensor, (electric) potential difference, temperature, pressure and force. These are connected to digital or analogue channels inputs on the interface.

Specially designed 'PASCO DataStudio' software is used to record the data into the computer and is available for Mac and Windows, although it has now been superseded by 'PASCO Capstone'.

Plug in the USB wire between the interface box and the computer. Select 'File, New Activity'. Then pick 'Create Experiment'. The interface should be detected and an 'Experiment Setup' window should appear. Plug the sensor into the interface box. Select the correct sensor from the list by double clicking on it: see Figure 35. For example, choose the photogate from the 'ScienceWorkshop Digital Sensors' drop down. Select the measurement required, for example 'Velocity in Gate, Ch1'.

Choose the method required for the display by clicking and drag-

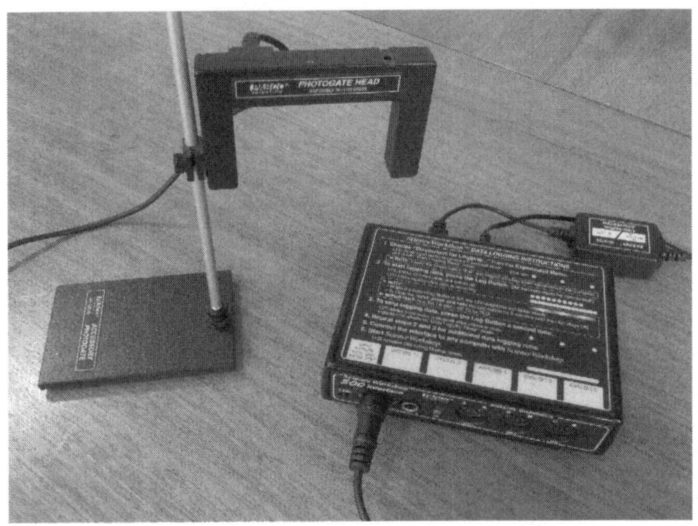

Figure 34: A PASCO Science WorkShop 500 Interface box connected to a photogate.

ging a display type from the 'Displays' panel at the bottom left to the measurement in the 'Data' panel at the top left. For example, choose 'Digits'.

To record data, press the 'Start' button in the menu bar at the top of the screen. The 'Digits' display will update each time the light beam is interrupted. Data taking can be stopped using the 'Stop' button. See Figure 36.

Alternatively, setup an ultrasound motion sensor to measure position and select a 'Graph' from the 'Displays' panel. Use the 'Motion Sensor' tab in the 'Experiment Setup' window to calibrate the sensor. Position the sensor a 'Standard Distance' from a large solid surface such as a wall, floor or a large book. Then press the 'Set Sensor Distance = Standard Distance' button. Record the data by using the 'Start' button as before. See Figure 37.

A table of values can be saved by dragging the 'Table' option from the 'Displays' panel up to the 'Data' panel even after the data has been recorded. The data can be saved for use later or in other software by

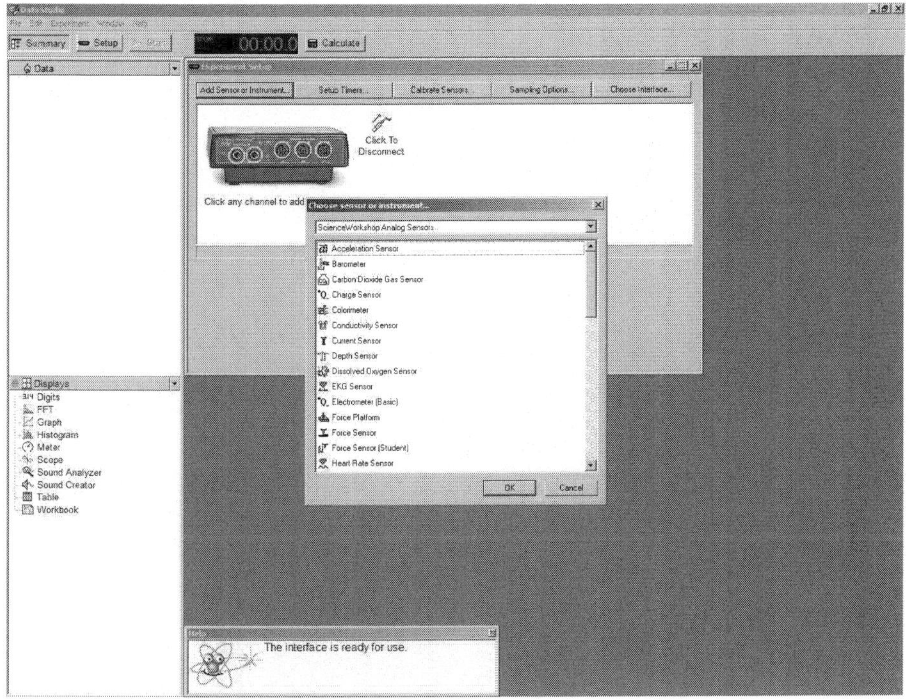

Figure 35: Choosing a sensor in PASCO DataStudio.

choosing 'Export Data' from the 'File' menu.

3.19 Data Harvest

Data Harvest QAdvanced dataloggers give reliable datalogging which is quick and easy to setup, see Figure 38. They can be used in either a standalone mode or when connected to a computer. When connected by a USB cable to a (Windows) computer the 'Data Harvest EasySense' software can be used to record, graph and save data. With a sensor connected to one of the inputs on the datalogger, when the programme opens choose 'EasyLog' and then press the 'Start/Stop' button. Live data will be displayed on the graph. See Figure 39. Alternatively, in a stand alone mode, the results are displayed on the built in screen or recorded and saved to the datalogger.

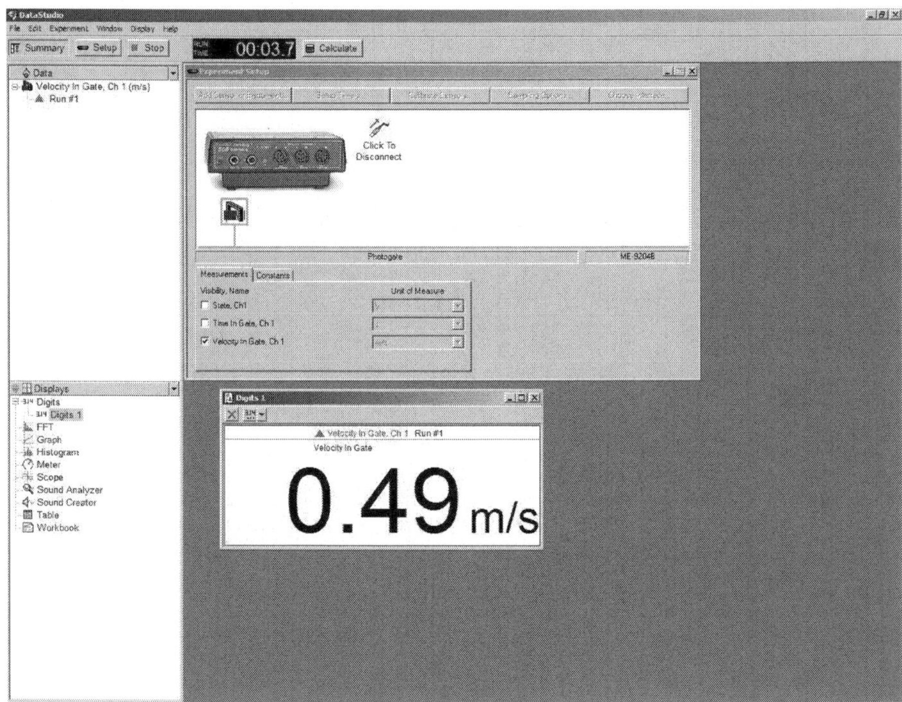

Figure 36: Using PASCO DataStudio to measure the velocity in a photogate.

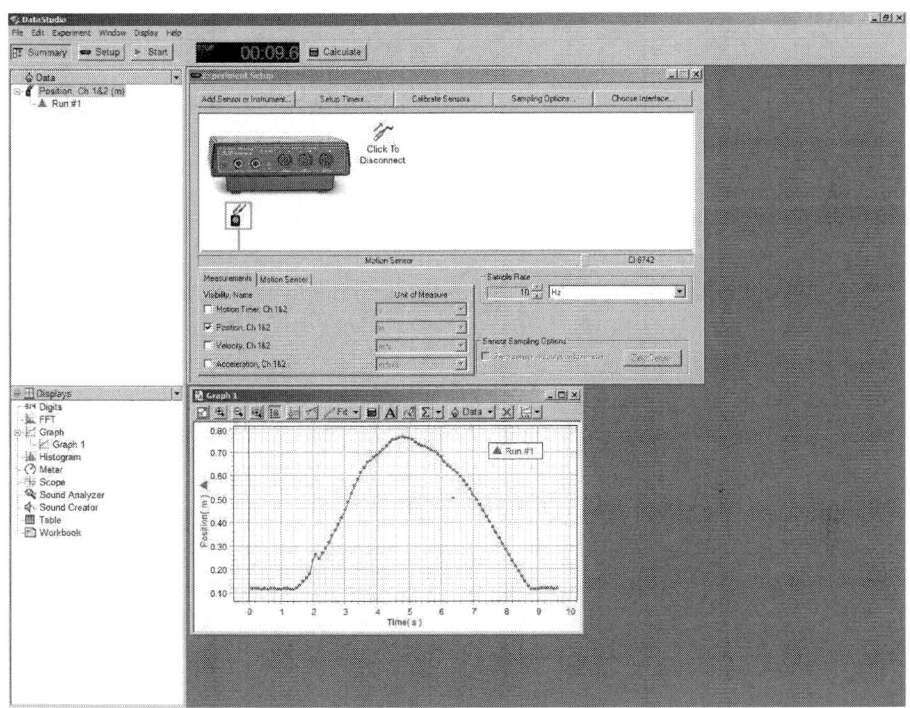

Figure 37: Using PASCO DataStudio to produce a distance time graph.

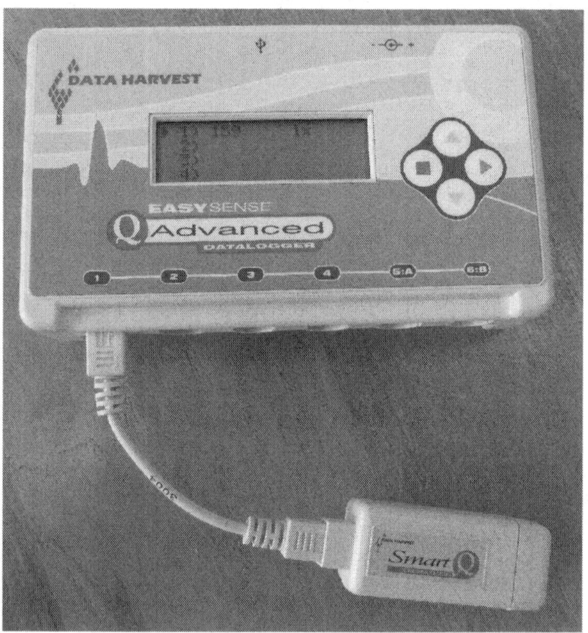

Figure 38: A Data Harvest QAdvanced Datalogger with a Light Level probe connected to input 1.

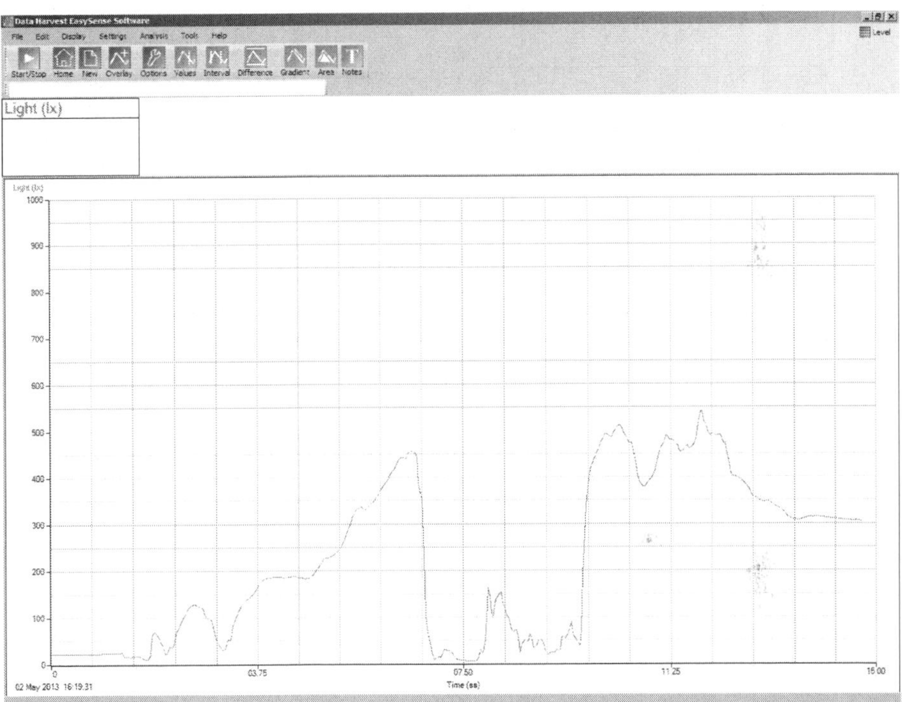

Figure 39: A screen shot of the Data Harvest EasySense software showing a graph of Light Level against time.

For example to setup a velocity measurement using a single light gate, connect the light gate to the input labeled A. Press any button to turn the device on. The arrow buttons can be used to navigate through different menu options. Navigate to 'Time & Motion' and press the green arrow button. Navigate to 'Speed' and press the green arrow button. Navigate to 'Speed at A' and press the green arrow button. Choose the '1 Interrupt card' option and press the green arrow button. Finally navigate to the length of the interrupt card being used and press the green button. The datalogger will display the latest speed as well as an average speed.

To use a Light Level probe, connect this to input 1. Navigate to 'Meter' and press the green arrow button. The display will show a reading updated a few times per second.

Use the red square button to go back to the previous menu. Navigate to 'Logging' and press the green arrow. Use the arrow keys set a 'Duration' and 'Interval' for the measurements and choose a start time. Logged data is saved to the datalogger, this can be downloaded to the 'Data Harvest EasySense' software on the computer for graphing or analysis. When the programme starts, choose 'Remote' and then 'Retrieve Remote'. Select the data to collect from the datalogger and click the 'Retrieve' button. The data is now displayed as a graph on the computer: see Figure 39.

3.20 Pressure Gauges

There are a number of pressure gauges which may be met. Each is based upon a different mechanical or electrical way of detecting pressure and is sometimes only suitable for certain pressure ranges or a certain resolution. A table of different pressure units was given in Section 2.5.1.

A basic pressure gauge can be made using a 'U' shaped tube which is half full of a liquid (such as water or mercury). One side of the U tube is connected to a known reference pressure P_1 such as atmospheric pressure or a vacuum and the other is connected to the pressure region to be measured P_2. The difference in liquid level, h, between the two sides of the 'U' tube is related to the pressure difference:

$$P_{1-2} = \rho g h \tag{21}$$

where g is the acceleration due to gravity and ρ is the density of the liquid.

Mechanical gauges measure pressure by monitoring an element which flexes or bends due to a pressure difference across it. The element which may be a Bourdon tube, a diaphragm or another similar device which causes a pointer to move against a calibrated scale. A Bourdon tube is a flattened tube which tries to regain its circular cross section when it is pressurised. This effect can be amplified sufficiently to turn a pointer by forming the tube into a 'C' shape or a coil. A diaphragm is a flexible membrane which separates areas of different pressures. The change in position of the membrane is reproducible, related to the pressure difference across the membrane and can be used to move a pointer.

The mechanical gauges may be read electronically by linking the Bourdon tube to a piezoelectric strain gauge or the diaphragm to a capacitance sensor.

A Pirani gauge is a common electronic gauge useful for measuring pressures between 0.1 Pa and 1000 Pa. It consists of a metal wire which is heated by a current flowing through it. Thermal equilibrium of the wire is achieved due to a cooling effect of the gas surrounding the wire. If the gas pressure decreases there will be fewer collisions of gas molecules with the wire meaning its temperature will increase. The increase in temperature of the wire causes its resistance to increase. Measurements of the current flowing through the wire and the electric potential difference across the wire are used to find its resistance. A calibration table or mechanism is then used to convert the resistance of the wire in to pressure.

Ionisation pressure gauges are more sensitive and measure pressures in the range 10^{-8} Pa to 10^{-1}Pa. Free electrons are generated which ionise gas molecules. These gas molecules are attracted to an oppositely charged electrode. The small current between the positive and negative electrodes is proportional to the rate of ionisation. Lower density (therefore lower pressure) gases produce less ions and so the current is linked to the pressure. However, the number of ions produced can be dependent on the gas so a good calibration for the appropriate gas is necessary. There are two subtypes: hot cathode gauges and cold cathode gauges. Hot cathode gauges use a heated filament to produce electrons via a process called thermionic emission. The most common of these is called a Bayard-Alpert gauge. A cold cathode gauge produces electrons by high emf discharge. The most common of these are called a Penning gauge or an Inverted Magnetron.

3.21 Temperature Measurement

There are a board range of methods for measuring temperature in the laboratory. The chosen method will be different depending on the temperature range to be measured, the accuracy and precision required and the conditions under which the measurement must be taken. Temperature can be measured in Kelvin (K), Celsius (°C) or Fahrenheit (°F). Temperatures can be converted from one scale to another using:

$$T_F = 1.8T_C + 32 \quad (22)$$

$$T_K = T_C + 273.15 \quad (23)$$

3.21.1 Near Room Temperature

Mercury and Alcohol Thermometers

A glass column is attached to a small reservoir or bulb at one end. This is filled with a liquid whose volume depends significantly on temperature. The two most common liquids are mercury and alcohol. Mercury thermometers are becoming rarer as government rules try to reduce the amount of mercury in the environment and as of 2012 their sale to (although not use by) the public is prohibited. Mercury thermometers have been replaced by coloured chemical thermometers. As the liquids increase in temperature their volume increases and they expand up the glass column.

Mercury thermometers are limited by the temperature range in which mercury is a liquid (-39°C up to 357°C). Similarly ethanol thermometers only work when ethanol is a liquid between -114°C and 78°C. To enable mercury free thermometers to be used above 78°C other chemicals can be used such as isoamyl acetate, toluene or kerosene which have higher boiling points.

IR Thermometers

An infrared thermometer infers the temperature of an object from the thermal radiation it emits. By knowing the amount of infrared radiation emitted by the object and its emissivity, the object's temperature can be determined.

A black body is an idealised object which absorbs all incident electromagnetic radiation no matter what the angle of incidence or frequency. A black body emits electromagnetic radiation, of intensity, I, where $I(f,T)$, is the amount of energy per unit surface area per unit time per unit solid angle emitted at a frequency, f, by a black body at temperature T. This is Planck's Law:

$$I(f,T) = \frac{2hf^3}{c^2}\frac{1}{e^{hf/kT}-1} \tag{24}$$

where c, is the speed of light, h is Planck's constant and k, is the Boltzmann constant. Integrating this over solid angle and all frequencies gives the Stefan-Boltzmann Law:

$$P = A\varepsilon\sigma T^4. \tag{25}$$

where P is the total power radiated by the object, A is the surface area of the object, ε is the emissivity of the object ($\varepsilon = 1$ for a perfect black body and $\varepsilon < 1$ for all other objects), σ is the Stephan-Boltazman constant which is $2\pi^5k^4/15c^2h^3 = 5.67 \times 10^{-8}\,\text{Wm}^{-2}\text{K}^{-4}$ and T is the temperature of the object. This means that the amount of emitted radiation is determined by only the temperature of the object.

Infrared thermometers are useful as they can make non-contact measurements in environments where other thermometers would not work or would be difficult to implement such as in a vacuum, at very high temperatures, in electromagnetic fields or where access is difficult. Sometimes they have a built in laser spot used to aim the thermometer indicating approximately from where the emitted radiation is being measured.

A infrared thermometer may give an inaccurate reading when it is not correctly aimed, when the emissivity is only given an approximate value for a particular surface/object or if stray radiation is picked up from nearby objects (usually at a higher temperature).

Thermocouples

Any conductor which is subjected to a thermal gradient (between the temperature of a sample and known reference temperature) will generate a small (electric) potential difference across its ends. This is known as the thermoelectric effect or Seebeck effect. It can be measured using a voltmeter.

In order to form a complete circuit to measure this (electric) potential difference a further conductor must be connected to the sample end of the first conductor. This further conductor will be subjected to the same temperature gradient which will develop its own (electric) potential difference which will oppose and exactly cancel the original (electric) potential difference. However, the magnitude of the effect depends on the metal from which each conductor is made. Thus, using two dissimilar metals results in a small, but measurable, net (electric) potential difference. The measured (electric) potential difference increases with temperature, and is typically between 1 and $70\,\mu\text{V}/^\circ\text{C}$.

Letter	Composition	Temp Range	Sensitivity
K	Ni-Cr	-200 to 1250°C	40 μV/°C
E	Ni/Cr-Cu/Ni	-50 to 740°C	68 μV/°C
J	Fe-Cu/Ni	-40 to 740°C	55 μV/°C
N	NiCrSi-NiSi	-270 to 1300°C	40 μV/°C
B	Pt/30 %Rh-Pt/6 %Rh	50 to 1800°C	10 μV/°C
R	Pt/13 %Rh-Pt	50 to 1600°C	10 μV/°C
S	Pt-Pt/10 %Rh	50 to 1600°C	10 μV/°C
T	Cu-Constantan	-200 to 350°C	40 μV/°C
C	W/5 %Re-W/26 %Re	0 to 2320°C	
M	Ni/Mo-Ni/Co	0 to 1400°C	

Table 13: Thermocouples, Compositions, Temperature Ranges and Sensitivities.

Since the (electric) potential difference is generated along the portion of the length of the two dissimilar metals that is subjected to a temperature gradient and both lengths of dissimilar metals experience the same temperature gradient, the end result is a measurement of the difference in temperature between the thermocouple junction and the known reference temperature.

Thermocouples can be purchased according to standard specifications denoted by letters. Different types are best suited for different applications. They are usually selected on the basis of the temperature range, sensitivity needed, how inert the metals are in the measurement environment and sensitivity to magnetic fields. Table 13 gives the properties of some common thermocouples: K is the most commonly used, and E is not affected magnetic.

3.21.2 Low Temperatures

A range of methods have been developed to allow measurement of low temperatures. Each method has associated advantages and disadvantages. It is likely that one or more of these methods will be met but the one in use will depend upon the experimental requirements and environment.

Gas Thermometry

With this method it is difficult to obtain a high accuracy due to the corrections needed for inaccuracies such as the volume of connecting tubing, contraction of the bulb and pressure corrections due to non-ideality of the gas. The pressure measurements become more accurate when diaphragms are used to separate the gas in the bulb and gas in the pressure gauge. Electronics can then be used to detect the movement in the diaphragm. This method also has the advantages that it is unaffected by magnetic fields and that it is easy to measure differential temperatures. Typically 3-He or 4-He gas would be used since they have a very low boiling point.

Resistance Thermometry

Resistance thermometry depends upon the variation of resistance with temperature. This method is much easier than the gas thermometry to perform and is reliable down to a lower temperature. A good material would have a (relatively) linear dependence of resistance on temperature, be easily obtainable in high purity, chemically inert and with a stable resistance. Metals such as platinum (Pt) are suitable. The possible resolution when using Pt decreases below about 10 K so materials such as Rhodium Iron (RhFe) are more useful. RhFe also has the advantage that is has a large and positive temperature coefficient below 30 K. Another useful material is Arsenic (As) doped Germanium (Ge). Thin Film Resistors using certain materials are prone to the effect of magnetic fields whereas carbon-glass thermometers are only slightly affected by a magnetic field. Thick film resistors using Ruthenium Oxide (RuO_2) have also been used due to their low cost, ease of use, ease of lowering their temperature due to a low heat capacity and the possibility of predicting their low temperature resistance from that at room temperature.

Electronic Thermometry

Electronic thermometry has a wide temperature range, a higher sensitivity compared to other methods, simpler operation and a more linear V verses T characteristic than most resistance thermometers. Silicone (Si) junctions make the most stable and reproducible thermometers, but they have high magnetic field dependence. A Gadolinium-Aluminium-Arsenic (GaAlAs) semiconductor has a much reduced dependence on magnetic field. They are inexpensive, but electrical noise limits currents to above $10\,\mu$A so heat dissipation becomes a problem at very low temperatures. Carbon and Germanium based semiconductors show a significant hysteresis effect when cooled again after being allowed to reach room temperature.

Thermocouples

Low temperatures can also successfully be measured using thermocouples. Stable, temperature sensitive alloys with a strongly temperature dependent magnetic moment must be used eg. those of Cu or Au. Great care is needed to get an accuracy of less than 1 % especially at low temperatures since the thermocouples become much less sensitive as the temperature drops.

Capacitance Thermometry

Capacitance thermometers, primarily using SiO_2-OH, have been successfully developed and have become popular due to their insensitivity to magnetic fields. The readings are stable and have a constant sensitivity right down to just above 0 K. However, after they are cycled to room temperature they show a significant hysteresis and drift effects.

Calibration

Thermometers operating at low temperatures can be calibrated in a number of ways. The first is by extrapolation of an existing scale down to lower temperatures. The second, is by direct measurement of a few known points such as in liquid helium (4.2 K), liquid nitrogen (77 K) and ice water (273 K). Mathematics can be used to interpolate temperatures between these. Alternatively a comparison with a primary (previously calibrated) thermometer can be made.

These methods can fail due to unknown additional temperature dependence at lower temperatures or due to unexpected and therefore uncorrectable systematic errors.

3.21.3 High Temperatures

High temperatures can be measured without contact using an infrared thermometer. Resistance thermometry may also be appropriate for contact measurements in a similar way as they are used at low temperatures. Platinum is particularly appropriate due to a linear and well defined resistance, its resistance to oxidation or other degradation in the high temperature environment and its relatively high melting point. A further method in common use is the thermocouple.

3.22 PID and Temperature Control

A PID controller is able to change and hold a parameter at a particular value called the setpoint. It consists of a control loop mechanism with feedback and is usually realised in an electronic circuit. It is commonly found in temperature controllers where a temperature can be set, monitored using a thermometer and increased using heater. Cooling usually occurs due to heat transfer to the surrounding.

A PID controller is named after its three correcting terms: the proportional, integral and derivative. These are summed to calculate the output of the PID controller. Defining $\mathrm{x}(t)$ as the controller output (which in the case of a temperature controller may be applied to the heater), the PID algorithm is:

$$\mathrm{x}(t) = Pe(t) + I\int_0^t e(\tau)\,d\tau + D\frac{d}{dt}e(t) \tag{26}$$

where P is the proportional gain constant, I is the integral gain constant, D is the derivative gain constant, e is the error (the difference between the set point and the present value), t is the time and τ is the variable of integration.

The proportional term produces an output value that is proportional to the current error. The proportional output is found by multiplying the error by the proportional gain constant, P. A high proportional gain results in a large change in the output, $x(t)$ for a given change in the error. A small proportional gain results in a small change to the output for the same change in error making the controller less responsive or less sensitive.

The integral term produces an output value that is proportional to both the magnitude of the error and the duration of the error. It is the sum of the instantaneous error over time and gives the accumulated offset that should have been corrected previously. The accumulated error is multiplied by the integral gain constant I to give the integral output. The integral term accelerates the movement of the process towards the setpoint. However, since the integral term responds to accumulated errors from the past, it can cause the present value to overshoot the setpoint value.

The derivative term produces an output value that is proportional to the slope of the error over time. Is is found by multiplying the rate of change of error with time by the derivative gain constant, D. The derivative term slows the rate of change of the controller output. Derivative control is used to reduce the magnitude of the overshoot produced by the integral component. However, the derivative term slows the response of the controller. Also, differentiation of a signal amplifies noise and thus this term in the controller is highly sensitive

Parameter	Rise Time	Overshoot	Settling Time	Error
P	decrease	increase	little change	decrease
I	decrease	increase	increase	decrease
D	little change	decrease	decrease	little change

Table 14: Effect of changing PID values.

Control Type	P	I	D
P	$P_U/2$	-	-
PI	$P_U/2.2$	$1.2P/T_U$	-
classic PID	$0.6P_U$	$2P/T_U$	$PT_U/8$
some overshoot	$P_U/3$	$2P/T_U$	$PT_U/3$
no overshoot	$0.2P_U$	$2P/T_U$	$PT_U/3$

Table 15: Relationships for setting PID values.

to noise in the error term.

In order for a system to reach its setpoint quickly the PID values must be adjusted until their optimum value is found. This can be done manually: the I and D values are set to zero. The P value is increased until the present value oscillates. P should then be set to half the value which caused the oscillation. The present value will then overshoot the setpoint. The I value can then be increased from zero until the offset is corrected in a reasonable time. Finally the D value can be increased so that any instability in the system is corrected in a reasonable time.

Table 14 gives the effect of changing one of the PID constants independently: An alternative to setting the PID gain constants manually is to follow one of a number of pre-defined algorithms for setting them such as the ZieglerNichols method [8]. As before the I and D gain constants are set to zero. The P gain constant is increased until it reaches the ultimate gain P_U at which the output of the loop oscillates with a constant amplitude. The values of P_U and T_U (where T_U is the time period of oscillation) are used to set the values of the PID gain constants using the relationships shown in Table 15 [9].

3.23 Kaye and Laby Tables

A vital part of any Physics Laboratory has been a book of physical constants and material properties. It has allowed physicists to look up quantities they need for data analysis or to check the accuracy of values they have measured. The standard book is Kaye and Laby Tables of Physical and Chemical Constants. This can be purchased as a printed book, but is also now freely available online at http://www.kayelaby.npl.co.uk/. This book and website give a wealth of information on a wide range of topics. Naming only three as examples: nuclear decay chains for every radio active isotope, the pressure dependence of the boiling point of organic compounds and the velocity of sound in a range of gases.

3.24 Microscopes

3.24.1 Optical Microscopes

Optical microscopes can be a valuable tool for enlarging small samples. The wavelength of visible light is in the range 400 nm – 700 nm, so distances smaller than this can not be resolved by an optical microscope, no matter how good the quality.

To view samples under an optical microscope a good light source is needed to illuminate the sample. Typically optical microscopes fall into two categories. Most common are transmission microscopes, which illuminate a sample primarily from beneath and rely on light transmitted through the sample. More useful in physics is a reflection microscope which usually have larger lenses to allow the collection of more light. They rely on the sample being brightly lit from above and collect light reflected from the sample.

Most microscopes will have position adjustable eyepieces which should be set to suit the user. Both the eyepiece and objective lens will have distinct magnifications, the total magnification is given by their product. Better quality microscopes will have a range of different

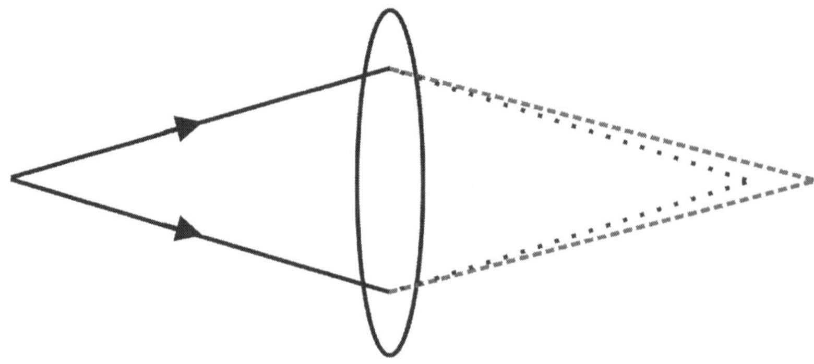

Figure 40: Illustration of chromatic aberation of a lens. Red light (the dashed line) is refracted less than blue light (the dotted line).

magnifications to choose from selecting the appropriate objective lens.

As well as the limit imposed by the wavelength of light, optical microscopes suffer from various aberations which reduce the quality of the images. Chromatic aberations occur when light of different colours (different wavelengths) is not focused at the same point. This is due to an effect called dispersion of the lens which arises because the lens has a decreasing refractive index for increasing wavelengths of light. Figure 40 illustrates the effect of a chromatic aberation. Chromatic aberation produces fringes of colour along boundaries between light and dark parts of the image. A simple method of reducing chromatic aberation is to make the focal length as long as possible or use monochromatic (a single wavelength) of light. More modern methods involve the use of low dispersion glass or the production of achromatic lenses using layers of materials with different dispersions.

Spherical aberations result from the use of spherical lenses. These are cheaper and easier to produce than aspherical lenses. Light which strikes the lens further from the center is refracted too much compared to light which strikes the lens nearer the center. Figure 41 illustrates the effect of spherical aberations. Spherical aberation can be reduced by making aspherical lenses or by combining a number of spherical lenses.

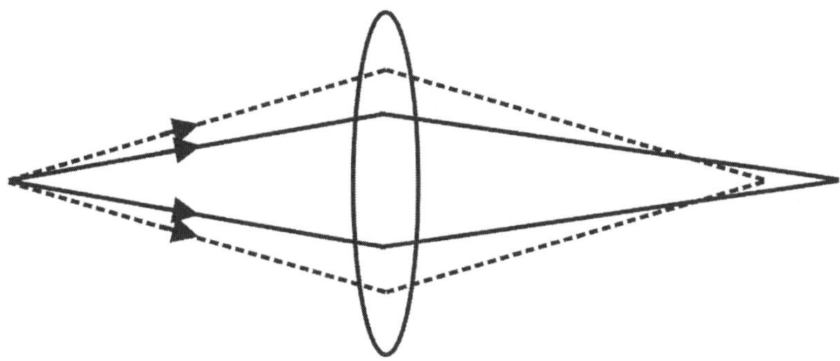

Figure 41: Illustration of spherical aberation of a lens.

Astigmatism occurs when light rays which propagate in two perpendicular planes have different foci. They typically arise as an artifact of the manufacturing process.

3.24.2 Transmission Electron Microscope

To observe distances smaller than the limit of an optical microscope electrons can be used as they have a shorter wavelength: typically 3.4 pm for 120 kV electrons. Samples must be electrically conductive. Insulating samples are first covered with a thin film of metal such as gold.

Electrons are usually created by thermionic emission and then accelerated by passing through a high potential difference. The higher the potential difference, the higher the energy and the smaller the wavelength of the electrons, meaning smaller features can be resolved.

The lenses in a transmission electron microscope (TEM) are made using electromagnets. Glass lenses would not work as electrons do not penetrate far into matter (hence very thin samples are needed to observe the diffracted beam). Permanent magnets would be equally useless because the image could not be focused. Electromagnets allow the lens currents to be varied and so allow the image to be focused onto a screen. A vacuum is needed inside the column of the microscope because electrons would collide with air molecules rapidly destroying

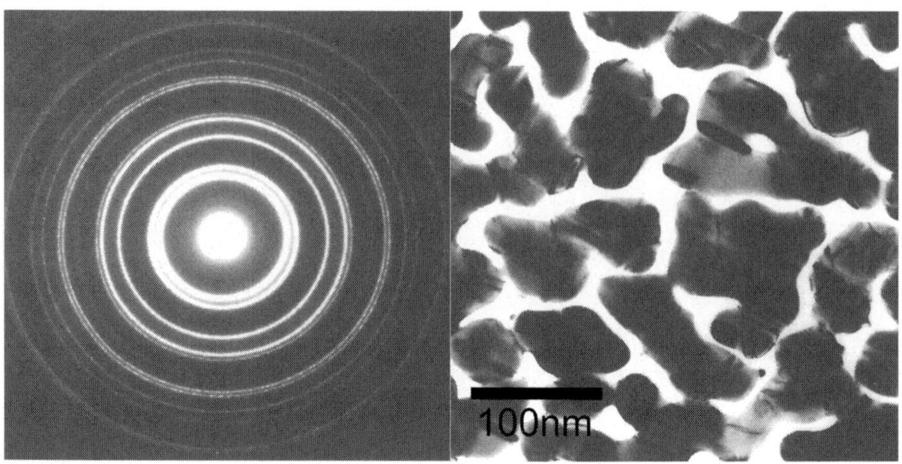

Figure 42: Left: Electron diffraction pattern from Gold. Right: A dark field image of islands of Gold.

the beam.

When the electron beam hits the sample some electrons are scattered and some are transmitted. If the transmitted beam is selected using the correct size aperture a bright field image is formed. An aperture is usually a platinum metal disk with a hole in the center placed in the path of the electrons. Since the scattered electrons are stopped by the objective aperture the amount of scattering determines the contrast in the bright field image. Where many electrons are scattered the image will look dark, where few are scattered the image will be lighter. A dark field image is formed when the scattered beam is selected using an aperture designed to block the direct beam. An advantage of dark field images is the much higher contrast which can be obtained, although a large electron flux is needed which may damage the sample.

3.24.3 Other Microscopes

Alternatively a scanning electron microscope (SEM), scanning tunnelling microscope (STM) or atomic force microscope (AFM) may be used. It is unlikely these microscopes will be encountered in the undergraduate laboratory, however it is still useful to have a brief overview of their operation. It should be noted that images taken with any of these microscopes are often given false colours. Since many of the features imaged are less than 400 nm (the wavelength of violet light) they can't actually have a colour in the traditional sense as they are not capable of reflecting visible light.

A scanning electron microscope (SEM) works in a similar way to a TEM except a narrow focused beam of electrons is scanned across the surface of the sample producing a image with a large depth of field giving a three dimensional appearance and a resolution down to about 1 nm. As with a transmission electron microscope, traditional SEM require samples to be placed in a high vacuum and be electrically conductive.

There are a range of methods of making measurements: transmitted electrons can be monitored or back scattered (reflected) electrons can be collected which allows areas with different chemical compositions to be identified as elements with high atomic numbers back scatter electrons more than elements with low atomic numbers. Alternatively characteristic X-rays can be detected, they are emitted when the electron beam removes an electron from the inner shell of an atom. This causes a higher energy electron to drop down to take the place of the ejected electron, releasing energy in the form of an X-ray. Characteristic X-rays are used to identify elements present and their relative proportions in the sample. Acceleration and scintillation based detection of secondary electrons allowed the development of an environmental scanning electron microscope which allows insulating samples to be imaged in low pressure gas or water vapour. Secondary electrons are those removed from the 1s (or k) orbital in atoms in the sample by inelastic scattering with electrons from the electron beam.

A scanning tunneling microscope (STM) can give a resolution of around 0.1 nm and a depth resolution of 0.01 nm. This is sufficient to image and even move individual atoms. Little sample preparation is needed: the STM works in vacuum, air or liquid over a wide range of temperatures. A very sharp conducting tip is positioned above the sample and scanned across the surface to build up the image pixel by pixel. A potential difference (or bias voltage) is applied between the sample and the tip which allows electrons to quantum mechanically tunnel between them. The resulting tunneling current can be measured and used to produce an image of the hight/relief of the sample. The challenge with STM is to have a very clean sample surface, the tip needs to be very sharp with just a single atom at the tip and there needs to be very good vibration suppression in the equipment and the building.

An atomic force microscope (AFM), similar to the STM, also has a sharp tip which is scanned across the surface of a sample. However with an AFM the tip is on the end of a cantilever. As the tip is brought close to the surface of the sample forces between the tip and the surface cause the cantilever to deflect a small amount. This is typically measured using a laser which is reflected from the top surface of the cantilever. This 'optical lever' allows a much bigger deflection to occur in the laser which can more easily be measured. Usually there is a feedback mechanism which adjusts the height of the sample to avoid the tip colliding with the surface. Images produced with an AFM give a three dimensional view of the surface of the sample. The sample needs no special treatment and can be imaged at room temperature in either air or liquid.

3.25 Spectroscopes

A spectroscope or spectrometer takes light from an object and splits it into its component colours/energies. Typically light will be brought to a focus on a slit at the entrance to the spectroscope. Light which passes through the slit falls onto a mirror which collimates the light i.e. makes the light rays parallel. This light is then directed to a diffraction grating. The angle at which the intensity minima and maxima of the interference/diffraction pattern occur depends on the wavelength of the light. The result is that the incident light is spread out. If white light entered the spectroscope the spectrum would be a rainbow pattern. If a sodium lamp is used, this only emits two specific wavelengths of orange light so there would be an emission spectrum showing two narrow orange lines against a dark background. A camera may be used to image the resulting spectrum and measurements may be taken to identify the wavelength of the emission or absorption lines.

3.26 Computers

3.26.1 Plugs, Sockets and Connectors

Table 16 gives images of a wide range of connectors which might be frequently used in the laboratory. It is does not constitute a complete list: even for a type of connector which is listed there may be other sizes or shapes of connector not shown. These images may be helpful in identifying unknown connectors.

Name	Male (Plugs)	Female (Sockets)
Mains		
Mains	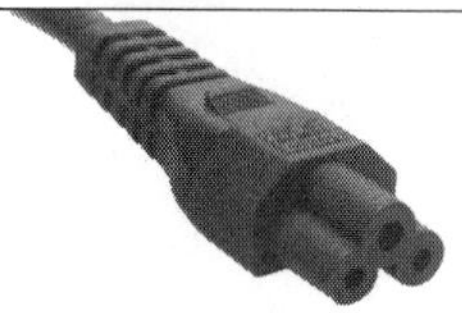	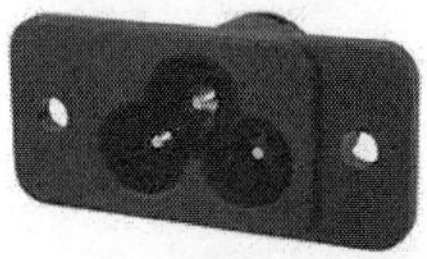
Mains		

3 Phase Mains		
VGA	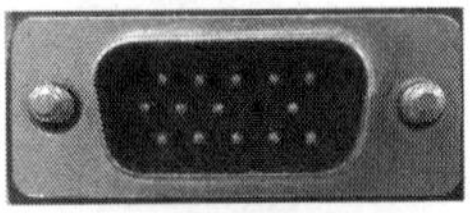	
DVI	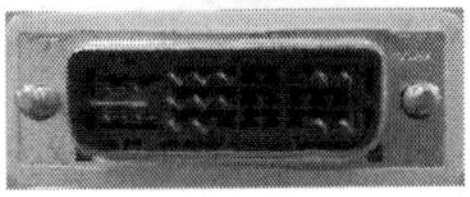	
S-Video	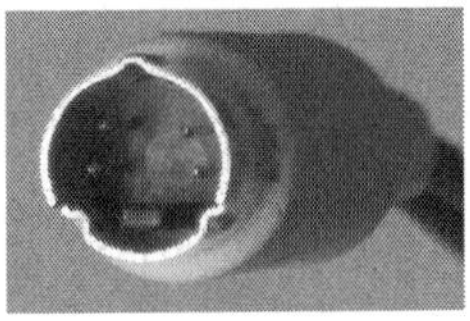	
Composite	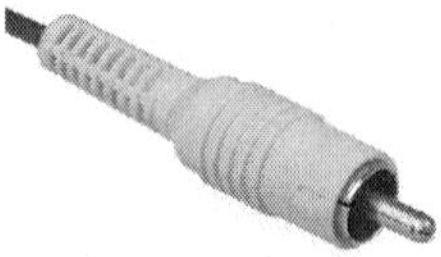	
HDMI		
3.5mm Audio		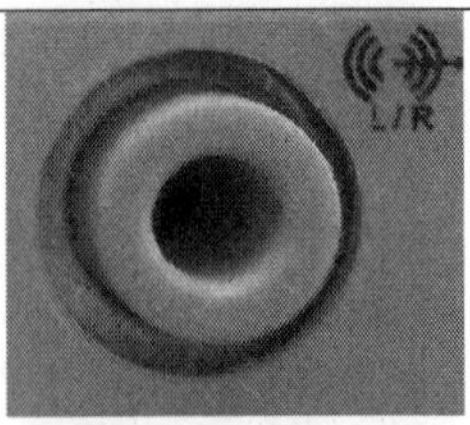
USB Type A	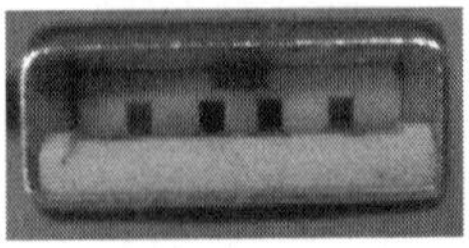	

USB Type B
mini USB
USB 3
miniUSB 3
FireWire (IEEE1394)
1394
mini FireWire
BNC

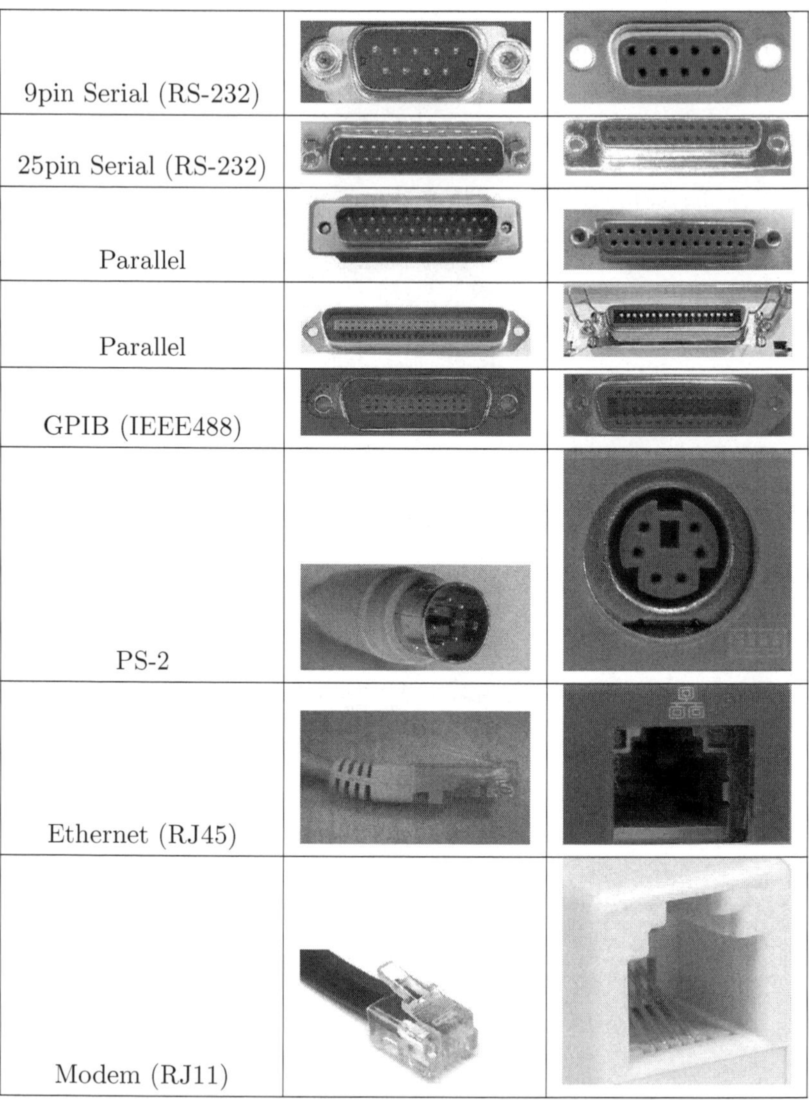

9pin Serial (RS-232)		
25pin Serial (RS-232)		
Parallel		
Parallel		
GPIB (IEEE488)		
PS-2		
Ethernet (RJ45)		
Modem (RJ11)		

Table 16: Common plugs and sockets.

3.26.2 Computer Data Logging

This section provides a brief tutorial on setting up Delphi, LabView and Matlab to record experimental data. It is likely that any computer used in the undergraduate laboratory will already be setup. However this section aims to give an overview of what is going on 'behind the scenes' and provide a starting point for setting a computer at a later point in time.

Instruments can communicate with a computer via a number of different connection interfaces. This simply involves connecting a cable from the instrument to a computer. Old interfaces include parallel and serial ports, but a more modern method is by USB. Frequently on modern computers serial ports are becoming obsolete although they can often be purchased on separate PCI cards or as a USB adapter.

More specialist scientific equipment often connects via ethernet (standard network cables) or GPIB (which requires a special adapter card inserted inside a computer or a USB converter).

To use a GPIB interface a GPIB PCI card (perhaps manufactured by Measurement Computing) needs to be installed into a standard Windows PC and the drivers correctly setup. A GPIB cable is run from the back of the card on the PC to the first instrument. Up to 14 further instruments can be connected in a daisy chain with a single cable linking an instrument to the previous one. Each instrument requires a unique address which is set in the configuration options on the instrument itself. (Addresses can be chosen regardless of where the instrument is located in the chain and also of which other address are chosen provided each is unique). The card driver installation on a Windows computer creates a shortcut called CBCONF32 in the Start Menu. This program is used to link the GPIB address of the instrument to a short text handle which is used to refer to the instrument in the programme code. Any other settings required can be configured using this program too, although the defaults are sufficient for most instruments.

Delphi

The Delphi programing environment needs to be installed on the PC. Each instrument requires specific text commands sent over the GPIB bus in order to respond with the required data. These sometimes complex commands can be written out once in Delphi in an interface unit specific to that instrument. Each Delphi program that is then written uses these interface units to communicate with instruments. A standard interface unit (called gpib) is also required in order for the Delphi program to communicate with the GPIB card in the computer. All the steps described in the this section need only be performed once. After this is all setup it provides a simple and rapid method to write new programs.

As an example consider reading the X output from a SR830 lock-in amplifier. A simple interface unit would be as follows below. The unit name (DCSTAN830) is given at the top, then there are two main sections 'interface' and 'implementation'. The interface section contains definitions of class properties, global variables and all procedures contained in the unit. The first line of the implementation section contains a list of other units from which code is referenced rather than duplicated. The code at the very end encapsulated with a 'begin' and 'end' statement creates the class and sets a text handle which corresponds to the text handle linked to the correct GPIB address set earlier in the CBCONF32 program. There are two procedures also in the implementation section, the first initialises the instrument and the second 'GetVolts' sends a message over the GPIB interface ('SNAP?1,2,3,4') and waits for the response to come back before interpreting it and saving part of it as the variable x. These messages and the form of the response are detailed for each instrument in its manual.

```
unit DCSTAN830;

interface

type  TLIA= Class
  ieeehdl:integer;
  ieeestr:String;
  x:real;
Procedure Initialise; Procedure GetVolts; end;

var LIA:array[1..4] of TLIA;

implementation

uses dieeecb,dmisc,sysutils,gpib;

Procedure TLIA.Initialise; begin
  ieeeHdl:=InitDevice(ieeeStr);
  ibclr(ieeeHdl);
  writeieee(ieeeHdl,'OUTX1');
  writeieee(ieeeHdl,'*CLS');
end;

Procedure TLIA.GetVolts; var s:string; begin
  writeieee(ieeeHdl,'SNAP?1,2,3,4');
  readieee(ieeeHdl,s);
  x:=s2r(copy(s,1,pos(',',s)-1));
end;
```

```
begin LIA[1]:=TLIA.Create; LIA[1].ieeeStr:='SR830'; end.
```

As well as the interface unit the actual program is needed – the simple one shown below sets a label on the Delphi form to show the lock-in X value when a button is pressed. At the top in the 'uses' section the name of the interface unit (DCSTAN830) has been added: this automatically includes the code from the file above to save it being typed out again. The 'FormCreate' function is run automatically when the program is opened and runs the instrument initialisation procedure. The 'Button1Click' function is run when the button on the form is clicked. This first runs the 'GetVolts' procedure to get the value from the instrument and then sets the label caption to be the value returned and saved to the variable x.

```
unit Unit1;

interface

uses
  Windows, Messages, SysUtils, Variants, Classes, Graphics, Controls, Forms,
  Dialogs, StdCtrls, dcstan830, dmisc;

type
  TForm1 = class(TForm)
    Button1: TButton;
    Label1: TLabel;
    procedure Button1Click(Sender: TObject);
  private
    { Private declarations }
  public
    { Public declarations }
  end;

var
  Form1: TForm1;

implementation

{\$R *.dfm}

procedure TForm1.Button1Click(Sender: TObject); begin
  lia[1].getvolts;
  label1.caption:=r2s(lia[1].x,13,-1);
end;
```

```
procedure TForm1.FormCreate(Sender: TObject); begin
  Lia[1].initialise;
end;

end.
```

This example can be extended to include all the instrument functions that are required. In practice the programs are run on a timed loop which reads the value on the instrument approximately every one second and records this to a data file or plots it to a live graph.

In order save the data to a file it needs to be assigned a handle using

```
AssignFile(handlename,filename);
```

It is then opened in append mode to add a line to the file or rewrite mode to over-write the current contents.

```
append(handlename);
```

or

```
rewrite(handlename);
```

The line is then written using

```
writeln(handlename, 'Text',_tb,VariableName);
```

where each element is separated with a comma, text is written in quote marks, variable values are written by including the variable name and a tab is written with '_tb'. The file is then closed using

```
closefile(handlename);
```

The 'handlename' variable needs to be declared as a textfile and the 'filename' variable declared as a string.

Matlab

Matlab can be used in a similar way to record data and plot graphs. The Instrument Control Toolbox must be installed. It comes with a comprehensive manual explaining how to communicate and control instruments via GPIB and Serial interfaces.

As a brief illustrative example, the following code snippet shows one of a number of methods data can be captured from an instrument connected via a GPIB interface. Here the data is stored in the variable 'data1'.

```
% Create a GPIB object.
obj1 = instrfind('Type', 'gpib', 'BoardIndex', 7, 'PrimaryAddress',
10, 'Tag', '');

% Create the GPIB object if it does not exist
% otherwise use the object that was found.
if isempty(obj1)
    obj1 = gpib('ni', 7, 10);
else
    fclose(obj1);
    obj1 = obj1(1)
end

% Connect to instrument object, obj1.
fopen(obj1);

% Communicating with instrument object, obj1.
data1 = query(obj1, 'SNAP?1,2,3,4');

% Disconnect from instrument object, obj1.
fclose(obj1);

% Clean up all objects.
delete(obj1);
```

LabVIEW

LabVIEW software gives a graphical point and click interface which allows acquisition and processing of data and data logging. It also has the ability to make decisions based on the measurements and control equipment. LabVIEW programmes can be run on Windows, Mac and Linux.

A LabVIEW program is called a VI. When LabVIEW is started, from the 'Create Project' window select the 'Blank VI' template: see Figure 43. This brings up two windows: one shows the 'block diagram' where the code is developed and the other shows the 'front panel' which is where the user interface can be customised to include graphs and buttons: see Figure 44.

In the 'block diagram' window, pictogram components can be wired together to create a programme. Each block has inputs and outputs just like a function in a traditional programming language. When a component executes it produces data which passes down the wire to the next block. The movement of data determines the order in which the components are executed in the programme.

The component from which the measurement is to be made needs to be

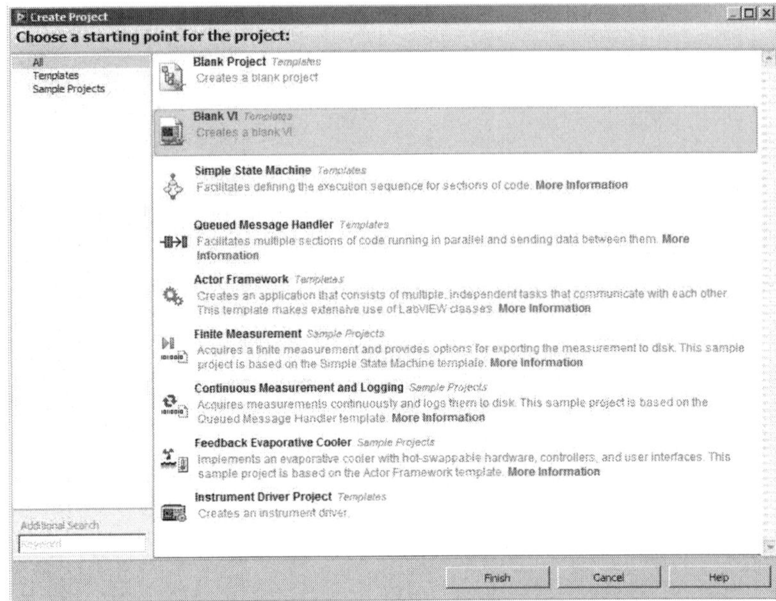

Figure 43: Creating a new VI in LabVIEW.

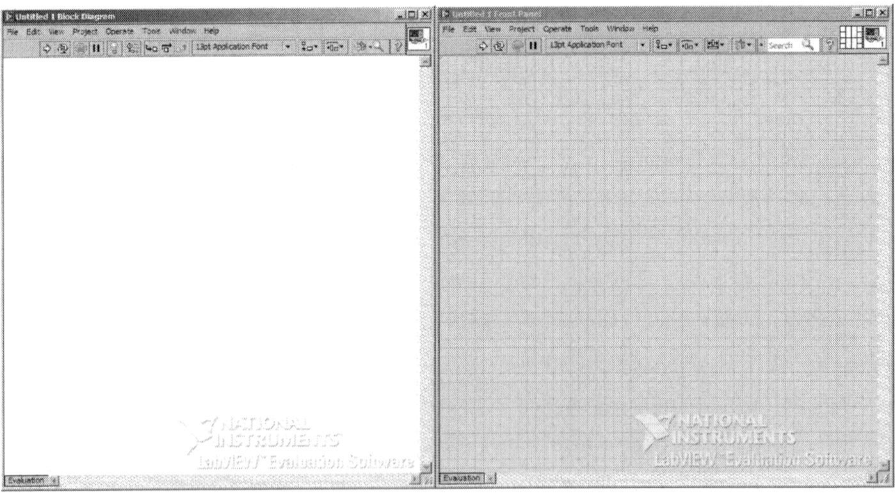

Figure 44: The LabVIEW programme. On the left is the 'block diagram' window and on the right is the 'front panel' window.

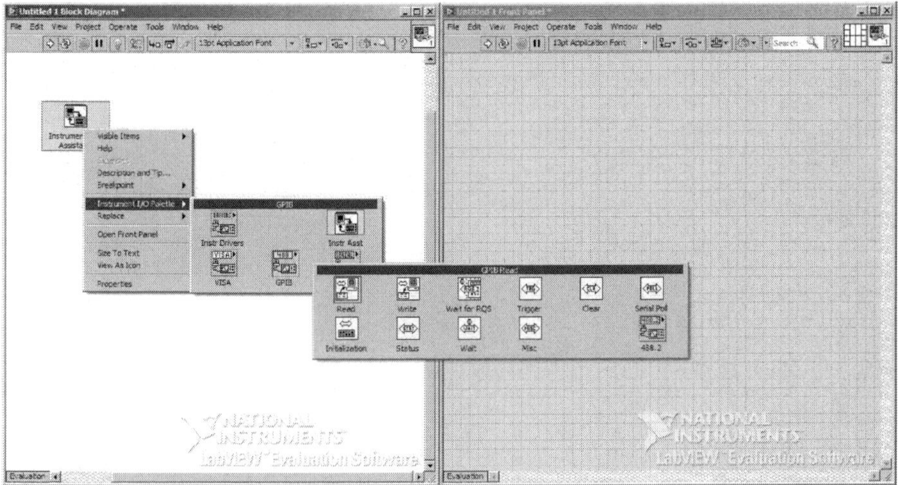

Figure 45: Inserting a 'GPIB Read' component in LabVIEW.

connected to the computer. A common way is via GPIB interface or via a USB system such as a National Instruments compact DAQ or an equivalent system made by Stanford Research Systems or another supplier. For example, to setup and read from the GPIB bus, right click in the 'block diagram' window and find the 'express' palette, then choose 'input' and the 'Instr Assist' button. Place the 'Instrument I/O Assistant' onto the block diagram. Right click on it and choose the 'Instrument I/O Palette', the 'GPIB' and finally the 'Read' button. See Figure 45. Place the 'GPIB Read' component onto the block diagram. Hovering the mouse over the different input and output markers around the edge shows that one is the 'address string'. Right click on this and choose 'String Palette' and then 'String Constant'. Place this on the block diagram and type in the address string for the instrument. The address string must be set to the text handle linked to the correct GPIB address via the CBCONF32 programme as discussed in Section 3.26.2. Also set a numerical value which indicates the number of bytes which must be read from the instrument.

Then wire the 'String Constant' component to the 'address string' input on the 'GPIB Read' component. The output from the 'GPIB Read' component is a string, this must be converted to a number for processing. Right click on the 'data' output marker of the 'GPIB Read' component and choose 'String Palette', then 'String/Number Conversion' and finally 'Fract/Exp String to

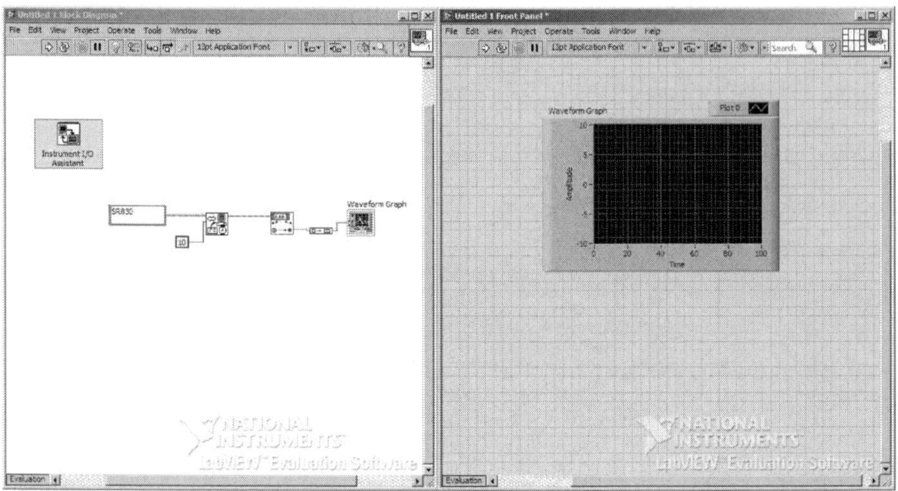

Figure 46: A LabVIEW programme which reads data and produces a graph.

Number'. Place this onto the block diagram and wire it up.

To set up a chart, right click in the 'front panel' window to bring up the 'controls' palette choose 'Graph Indicators' and then 'Waveform Graph'. Place this on the 'front panel' window. Now right click on the input to the 'Waveform Chart' component and select the 'Array Palette' option and then the 'Build Array' component. Add this to the block diagram. Now wire the output of the 'Fract/Exp String to Number' component into the 'Build Array' component and finally into the 'Waveform Graph'. See Figure 46.

This will only produce a single value. Inserting a 'while' loop will run the programme continuously until a 'stop' button is pressed. Right click in the block diagram window. Choose 'Exec Control' and then a 'While Loop'. Use the mouse to draw a box around all the components that are required to be inside the while loop. A stop button has also appeared on the 'front panel' window. See Figure 47. The programme can now be run, by pressing the run arrow in the menu bar at the top of either the 'block diagram' or 'front panel' windows. Live data should appear on the graph.

LabVIEW programmes can be more complex than this if more instruments are used and the data is written to a file using the 'Write Measurement File' component from the 'Express, Output' palette. Programmes for use by many users would normally be compiled as an executable so that changes to

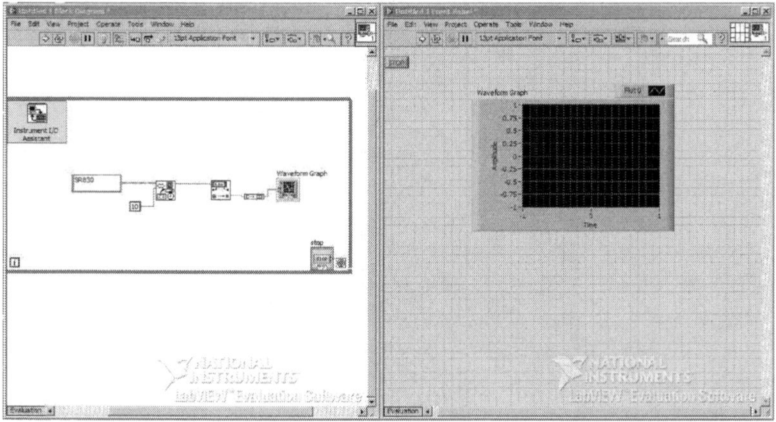

Figure 47: A LabVIEW programme within a 'while' loop.

the code can't be made by mistake.

3.26.3 Still Cameras and Scanners

CCD sensors (or digital cameras) are a useful way to capture data in am image format such as from diffraction/interference patterns, emission/absorption spectra and with the correct optics from a microscope. The CCD sensors used may be sensitive to a wide range of wavelengths of electromagnetic radiation such as X-rays, visible or infrared. A computer can then be used to make detailed measurements from the images using special software.

Often the image on the computer screen will need calibrating so that a certain on screen distance (or number of pixels) can be matched up with a certain physical distance in the real world. Typically a ruler or graticule (a very finely spaced series of lines used under an optical microscope) is photographed under the same conditions of magnification and focus as the sample. A special software programme can be used to calculate a scale and to make measurements from the image of the sample using the image of the graticule.

Alternatively, for a small number of measurements, a simple image editing programme can be used to measure the number of pixels between two graticule lines. Figure 48 shows a photograph of a graticule taken down a microscope. Each small division corrections to 100 μm. If this was measured to be 40 pixels, then each pixel would correspond to $100/40 = 2.5$ μm. Distances on the image

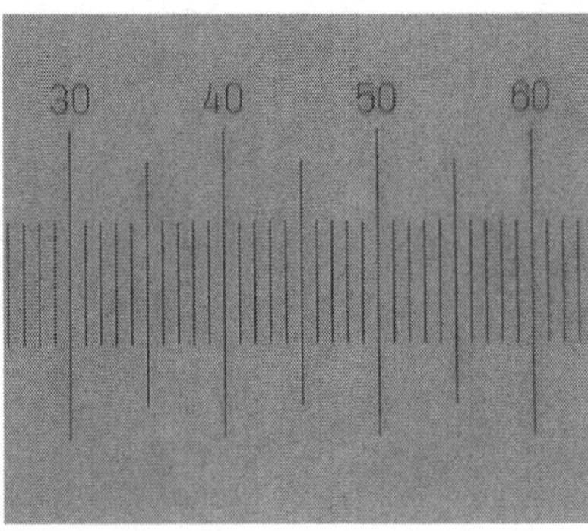

Figure 48: Photograph of a graticule using a microscope.

of the sample could then calculated by measuring the appropriate number of pixels and multiplying by 2.5 μm.

A similar procedure can be followed if images are scanned into a computer. This is perhaps most likely with images taken using traditional film and developed on photographic paper.

3.27 Video Cameras

Video cameras, camcorders and slow motion (high frame rate) video cameras can be used to record experimental data. They are particularly useful for measuring the speed of objects which are moving or events which happen too rapidly to be seen with the human eye. The experiment should be setup so that a ruler or scale is in camera shot and in close proximity to the object. Alternatively a resolution calibration must be performed giving a value in units of metres per pixel.

Captured video can be opened in a programme such as Apple QuickTime. The video should be played back frame by frame. The position of the object can be read from the ruler or by measuring its position in terms of pixels from a fixed point. When the same measurements are taken from the next frame the velocity can be calculated using:

$$v = \frac{\text{change in distance (m)}}{\text{change in time}} \tag{27}$$

or

$$v = \frac{\text{change in pixels (px)} \times \text{resolution (m/pixel)}}{\text{change in time}} \tag{28}$$

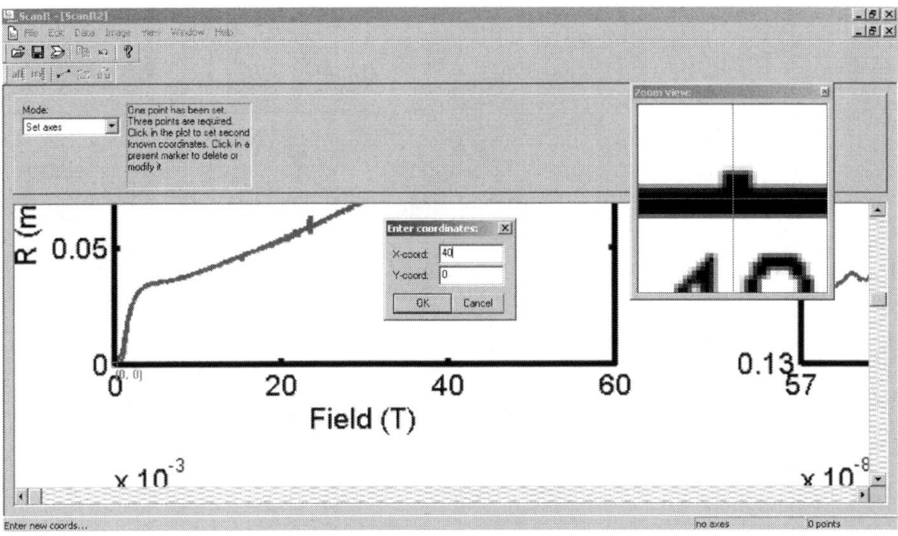

Figure 49: Using ScanIt to set the axis.

The change in time is obtainable from the software or by calculating one over the number of frames per second.

If using QuickTime, once the video file is opened, choose 'Show Movie Inspector' from the 'Window' menu. This window has, amongst other details, the 'Current Time'. The video can be moved one frame at a time using the right or left arrow keys on the keyboard.

3.28 Other Researcher's Data

It is frequently very useful to be able to take graphical data published in textbooks or other authors and to be able to replot this in a different way or in conjunction with your data. Research papers only rarely provide tables with the actual data points which make up the graphs, so more creative means must be adopted to extract the data.

A very useful programme called ScanIt [10] allows a scanned image of a graph to be loaded. Firstly, set the axes by clicking on three well defined points such as the origin, a point on the x axis and a point on the y axis. For each point input the coordinates based on the scale printed on the graph. This is shown in Figure 49.

The programme allows the user to either identify points manually or to automatically detect and trace a curve. Both methods produce a table of x and y coordinates which can be exported to a graph drawing programme.

3.29 Wii Controllers

Over recent year there has been great interest in using Nintendo Wii games console controllers as data loggers. Wii controllers have a built in 3 axis accelerometer and an infra red camera which can be used to give the distance from a pair of infra red sources. They can communicate with computers via their built in bluetooth radio.

Early work [11],[12] involved a significant amount of computer programming. Dedicated and easy to use 'Wiimote' software has now been written [13],[14] to allow the connection of Wii controller with a computer and subsequent collection of data in real time.

The computer must have a Bluetooth radio adapter. Sometimes these are built into computers (especially laptops), but a cheap USB dongle can also be purchased. The more stable and reliable ones tend to be those which use the Microsoft Bluetooth Stack drivers. Once the computer is setup, the Wii controller needs to be paired by pressing buttons 1 and 2 together and initiating the pairing via the computer. If a passcode is requested in Windows it can be left blank or skipped by pressing Alt-S on the computer keyboard.

Once the Wii controller is paired, the Wiimote software can be started. In accelerometer mode, the acceleration can be recorded in the x, y and z axis. It measures in units of g between $\pm 3g$ with a resolution of $0.04\,g$. When the controller is stationary lying flat on the desk, one of the axes will measure $1\,g$. This can be used to measure the acceleration of a dynamics trolley, a pendulum, a mass spring system, a lift, a car or a rotating disk.

The software can also be set in 'position sensor' mode. The bar which usually sits on top of the TV is commonly referred to as the 'Sensor Bar'. In fact, it does not sense anything, but is two infrared LEDs positioned 20cm apart. These two sources emit infrared light which is detected by the 1024 pixel by 768 pixel infrared camera in the Wii controller. Given the angular fields of view of the camera are 41° horizontally and 31° vertically it is possible to find the average angle subtended per pixel as 0.040°. The software in the Wii is able to measure the number of pixels, n, between the two light sources on its infrared camera and convert this into the angular separation. Since the sources are known to be 20 cm apart, the distance, d, from the WiiMote to the Sensor Bar can be calculated using triangulation:

$$d = \frac{0.1}{\tan(0.020 \times n)} \tag{29}$$

3.30 Traditional Photographs

Photographic film may still be used to take images from electron microscopes, X-ray cameras or to make holograms. Where even possible, CCDs for these operations are still expensive and sometimes difficult to fit to older equipment which is more likely to be present in undergraduate laboratories. Thus it is useful to know how to develop film and then to produce paper photographs.

The film is usually a piece of plastic with a silver halide (typically a combination of AgBr and AgI) emulsion coating and a protective gelatin layer. When the film is exposed to photons of electromagnetic radiation an electron within the halide ion absorbs the photon, gaining energy. This frees the electron so that it is able to move within the silver halide crystal. After some time, it can become trapped by a defect in the crystal (often a silver sulphide impurity) known as a sensitivity speck. The electron can then combine with a silver ion, forming a silver atom. The silver atom is an additional impurity which traps further electrons, creating further silver atoms.

Once exposed in the experiment, individual film packets should only be opened in a dark room. The dark room may be dimly lit with a coloured light compatible with the film (meaning the film has a low sensitivity to that specific colour of light) or in complete darkness. Usually the solutions necessary are placed in 4 side by side trays.

The film is first put into a developer solution. The developer is an alkali which which preferentially reacts with the silver deposits. Since it is a reducing agent it converts further Ag^+ ions into deposits of silver. This enhances the process which was started by the free electrons generated by the electromagnetic radiation.

If the film is left in the developer for too long, then all the silver halide crystals will be converted into silver atoms giving a black film. A brief wash in water can remove all the developer and prevents any further reactions.

The film is then placed into a third solution, the fixer. This is usually an acid solution which removes any silver halide crystals which have not reacted. Finally the film is washed again to remove the fixer which prevents the film turning brown due to the formation of silver sulphide. The films should then by hung up to dry or put through a suitable dryer.

The exact chemicals used for the developer and fixer depends on the film. They should be regularly changed if a large number of films are being developed. The time the film is left in the developer and fixer solutions has a large effect on the final image and will depend on their strength. The correct timings will usually be given, however as a guide consider around 1 minute in the developer, 30 seconds in the first wash, 2 minutes in the fixer and 2 minutes in the final wash.

To create a paper photograph a second stage of development is required. The developed film is inserted into a slot below the light source on an enlarger. The light shining through the film creates an image on the easel. Once the size and position of the image have been adjusted the light on the enlarger is turned off. A sheet of photographic paper is placed onto the easel. The light source can then be turned on using a timer, exposing the photographic paper to the image. The final stage is to develop the photographic paper in a similar way to the development of the film. The developer solution is usually specific to the paper, but the same two wash baths and fixer bath as for the negative can usually be used.

Always carefully mop up any chemicals which are spilt and clear away all chemicals at the end of each session/day as they can leave residues behind when they evaporate. Chemicals should not be returned to the original bottle as this will cause contamination.

4 Data Analysis and Errors

Perhaps at school, once a measurement was completed and a result calculated it was thought the experiment was complete. Sometimes the result may have even been compared with an accepted value. For instance, an experiment might have been performed to measure the acceleration due to gravity, g. Perhaps the oscillations of a simple pendulum were timed and the result substituted into the familiar equation:

$$T = 2\pi\sqrt{\frac{l}{g}} \tag{30}$$

At university level, experiments must be taken further to include an estimate of the error in the result. Scientists are usually interested in making measurements, not for their own sake, but to test a theory. Theories can be accepted or rejected on the basis of experimental measurements, therefore it is vital to have some indication of the accuracy of the results.

Every time a measurement is made there will be a degree of uncertainty in the result: an instrument may be difficult to read, it may fluctuate, it may be badly calibrated, a genuine mistake may be made reading the instrument or the result could be limited by the precision of the instrument. It is vital that a competent scientist has an understanding of the terminology of errors as well as how to calculate, evaluate and develop techniques to minimise the significance of measurement error in their work. It is beneficial to adopt an inquisitive mind-set where questions are always asked and improvements to the apparatus and techniques which may reduce measurement error are always considered: this can be summarised as 'be awkward'!

Suppose a large underground cavern is to be found. The hypothesis might be that due to the void in the ground the measured value of g near the cavern will be less than $9.81\,\mathrm{m/s^2}$. Of course, the familiar value of g=$9.81\,\mathrm{m/s^2}$ is not exact and so also has an error which is neglected in this illustrative example.

A value of $9.75\,\mathrm{m/s^2}$ is measured. Does this provide evidence of a discrepancy which may indicate the presence of the underground cavern? To find out an indication of the error in this measurement is needed. The result may be found to be $9.75 \pm 0.01\,\mathrm{m/s^2}$. This is inconsistent with with $9.81\,\mathrm{m/s^2}$ and so providing there is confidence that the experiment was well performed and a good estimate of the error made, there is evidence for the presence of the underground cavern.

Suppose the result is found to be $9.75 \pm 0.10\,\text{m/s}^2$. This time the magnitude of the error is a factor of 10 times larger. This means that the result is, in fact, consistent with $9.81\,\text{m/s}^2$ and that the result is in agreement with what is expected. The search for the cavern should continue elsewhere.

Finally, suppose the result is found to be $9.75 \pm 5\text{m/s}^2$. This result is also consistent with $9.81\,\text{m/s}^2$, however the error is quite large and means that even quite significant difference can not be detected. This raises an important point: it may not actually be possible to detect the cavern at all if the error in the results is too large. Frequently in the history of science, discoveries have been made and theories accepted or rejected by scientists who have developed their experiment and equipment to have a smaller error than anybody had previously. *If you want to make a discovery, as emphasised earlier in this chapter you should 'be awkward'!*

This chapter begins with an explanation of the terminology of errors. It then develops data and error analysis techniques with a particular focus on graphs.

4.1 Accuracy and Precision

The accuracy of a measurement is how far the measurement is away from the true value of the quantity that is being measured. The precision of a measurement is the full width of the spread of data measured when taking repeat measurements under identical conditions.

A measurement can be accurate and not precise, precise and not accurate, precise and accurate or neither precise nor accurate. Often either a target board (see Figure 50) or a distribution graph (see Figure 51) is used to illustrate this.

For example a measurement may have a large systematic error. This would mean that it was not accurate as the measurements would lie a long way from the true value. However, the measurement could be very precise as it could be repeated many times yielding the same result. Thus the results may be closely scattered but around the wrong value. A crude example would be measuring the length of a desk to the nearest centimeter using a meter rule on which the first 5 cm was missing. Alternatively a measurement may be taken with an instrument which has large fluctuations in the readings. This would mean that the measurement may be accurate, but not precise.

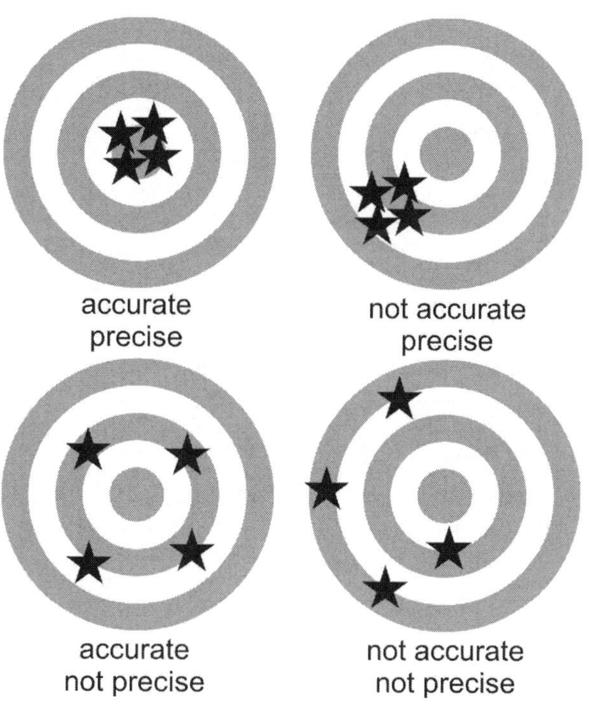

Figure 50: Targets illustrating the differences between the terms accurate and precise. The center of the target corresponds to the true value. The stars are the individual measurements.

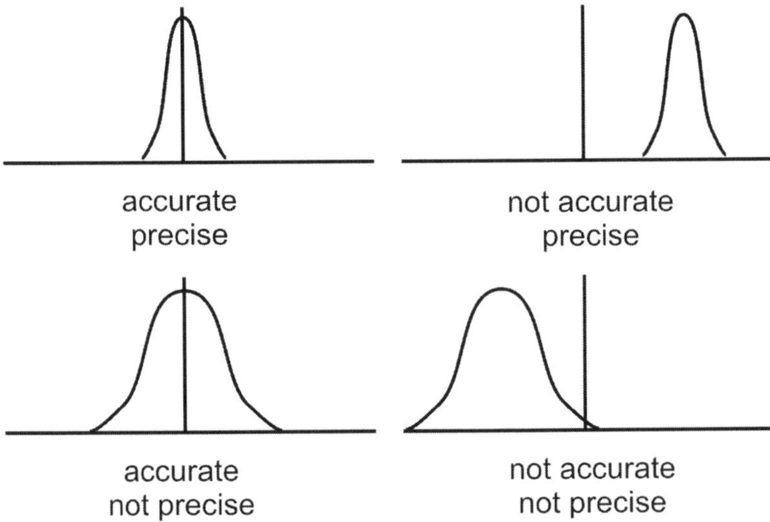

Figure 51: Graphs illustrating the differences between the terms accurate and precise. The vertical line corresponds to the true value. The vertical axis is a probability distribution (or the number of measurements) and horizontal axis is the magnitude of the quantity being measured.

4.2 Measurement Errors

Measurement errors can be split up and considered as having one of two distinct origins: they are either random errors or systematic errors. As will be seen, this distinction is important when considering the significance of a new result or when trying to reduce the measurement error.

4.2.1 Random Error

Random error, as the name suggests, is inherently unpredictable. It causes the results to be scattered around the true value. It arises from: natural fluctuations in readings or the environment; or the experimenters' interpretation of the readings. The mean of the fluctuations around the true value of a quantity has an expectation value of zero. Random error is linked to the term precision which is discussed above: a large random error indicates an imprecise result, conversely a small random error indicates a precise result.

4.2.2 Systematic Error

Systematic error is an error caused by the experimental technique or apparatus. It causes to the mean of the measurements to be different from the true value. It is important to estimate the size of systematic error as well as its direction. A systematic error may cause a result to be larger than the true value or less than the true value.

4.2.3 Zero Error

Zero error is a special name given to a common type of systematic error. This occurs when a measurement device does not read zero before being put into use. For example an analogue voltmeter may already read 1 V before being connected across a battery to measure its potential difference. Alternatively, a balance may read 3 g before a mass is placed upon it. It is relatively simple to account for zero errors, by making the relevant correction to each result. For example subtracting 1 V or 3 g from each measurement.

4.3 Uncertainty

Uncertainty is the degree to which a result is unknown. In the case of direct measurements of a quantity the uncertainty is half the resolution. It should be quoted as a plus or minus value on the given measurement. For example: the circumference of a can of beer could be 18.0 cm $\pm$ 0.5 cm. In more complex cases involving detailed analysis and non linear dependencies the plus error and minus error might be different. For example: the current best estimate of the mass of the up quark is $2.3^{+0.7}_{-0.5}$ MeV [15].

4.4 Resolution

The resolution of an instrument is the smallest change that the device can detect in the quantity it is measuring. It is the smallest scale division on the meter. For example a meter rule with markings only every 1 cm has a resolution of 1 cm. Whereas a meter rule with markings every 1 mm has a resolution of 1 mm. A digital stop watch giving a reading to two decimal places of a second (i.e. 1.22 s) has a resolution of 0.01 s.

4.5 Tolerance

The term tolerance is usually used to express the uncertainty that the design allows for during manufacture of an object or component. For example a piece of metal might be needed that is 1.000 m long with a tolerance of 1 mm. This means that the metal needs to be cut somewhere between 0.999 m and 1.001 m. Another frequently occurring example is a resistor. One of the colour bands on the resistor corresponds to the tolerance. For example a 100 Ω resistor with a gold band (which indicates a tolerance of $\pm$ 5 %) should have a resistance between 95 Ω and 105 Ω.

4.6 Sensitivity

In situations (eg. using an oscilloscope) where a quantity (eg. potential difference) is not displayed or measured with the same dimensions as the measurement (eg. cm rather than V) it is appropriate to define the sensitivity of a measuring device. This is the ratio of the change of output to the change of input. A high sensitivity would mean a large distance in cm corresponding to a small (electric) potential difference whereas a low sensitivity would mean a small change in distance corresponding to a large (electric) potential difference. In this example the sensitivity could be given as 2 V/cm meaning the oscilloscope screen gives a reading where 1 cm corresponds to 2 V.

4.7 Response Time

The response time is the time taken for the measurement device to respond and reach its final reading after a dramatic change in input. Digital meters typically have a faster response time than an analogue meter with a pointer because the analogue meter needs a comparatively long time to physically move. Changes in the quantity being measured which happen faster than the response time of the measurement device will be averaged out and not seen. A good demonstration of this would be using an alternating current from a signal generator as the input to a center-zero analogue meter. Starting from a low frequency (with period much longer than the response time) where the pointer is seen to oscillate the frequency can be gradually increased until the period is much shorter than the response time where the pointer will simply twitch slightly around a reading of zero.

4.8 The Mean of the Sample

Frequently in Physics measurements are repeated over and over again. It is probably easy to recall school teachers impressing the need to always repeat measurements. Why did they do this? Repeated measurements which agree or are consistent with each other increase the confidence that a precise result has been achieved.

When repeated measurements are made of a quantity subject to random error a mean can be calculated. If there are N readings (i.e. $y_1, y_2 ... y_N$) the mean can be found by adding up all the results and dividing the total by the number of measurements made. This is written mathematically as:

$$\mu = \overline{y} = \frac{\sum_{i=1}^{N} y_i}{N} \tag{31}$$

This assumes that any error of a particular size is equally likely to be positive as negative.

For example, suppose that a measurement of the current flowing through a lamp is measured 10 times and the values were:

$$0.563\,A, 0.497\,A, 0.501\,A, 0.489\,A, 0.512\,A,$$
$$0.492\,A, 0.536\,A, 0.472\,A, 0.520\,A, 0.484\,A$$

then using Equation 31, $\overline{y} = 0.507$. Taking the final result as the mean of

i	y_i	$(y_i - \overline{y})$	$(y_i - \overline{y})^2$
1	245	-4.5	20.25
2	235	-14.5	210.25
3	267	17.5	306.25
4	254	4.5	20.25
5	250	0.5	0.25
6	246	-3.5	12.25
Sum	1497		569.5
Mean	249.5		$s^2 = 94.92$
			$s = 9.74$

Table 17: Example data for calculation of the mean and variance.

a large number of readings rather than any one individual reading is useful as it reduces the effect of random errors. However it does not give the whole picture. It does not give any indication of the spread of the values around the mean. The smaller the spread the more likely the mean is to be equal (or very close) to the true value, providing there is no systematic error. A small spread means a high degree of confidence can be placed in the result.

4.9 Variance and Standard Deviation of the Sample

As a way to quantify the spread in a set of measurements the variance can be calculated. The variance, s^2, of a set of N measurements of the quantity y is given by:

$$s^2 = \frac{1}{N}\sum_{i=1}^{N}(y_i - \overline{y})^2 \tag{32}$$

where $\overline{y}$ is the mean of the set of N measurements given by Equation 31. s^2 will often be called the variance *of the sample* and s the standard deviation *of the sample.* The standard deviation is the square root of the variance. A large variance indicates a large spread in the measurements and visa versa for a small variance. As an example take the data shown in Table 17.

Equation 32 might look over complex as the deviation of a single measurement from the mean is given by $(y_i - \overline{y})$ and that it would be possible to find the variance from $s^2 = \frac{1}{N}\sum_{i=1}^{N}(y_i - \overline{y})$. However, this does not work in practice because $\sum_{i=1}^{N}(y_i - \overline{y})$ is always exactly equal to zero as a result of the definition of the mean.

It is valuable, especially when writing computer code, to consider that Equation 32 for the variance can be rewritten:

$$s^2 = \frac{1}{N}\sum_{i=1}^{N} y_i^2 - 2\overline{y}y_i + \overline{y}^2 \quad (33)$$

$$= \sum_{i=1}^{N} \frac{x_i^2}{N} - \frac{2\overline{y}y_i}{N} + \frac{\overline{y}^2}{N} \quad (34)$$

$$= \overline{y^2} - 2\overline{y}^2 + \frac{\overline{y}^2 N}{N} \quad (35)$$

$$= \overline{y^2} - \overline{y}^2 \quad (36)$$

where

$$\overline{y^2} = \sum_{i=1}^{N} \frac{y_i^2}{N} \quad (37)$$

and

$$\overline{y} = \sum_{i=1}^{N} \frac{y_i}{N} \quad (38)$$

This has significant advantages when computing the variance of a large data set on a computer as only one pass over the data is required. Both $\overline{y}$ and $\overline{y^2}$ can be calculated in one pass, rather than requiring a first pass to evaluate $\overline{y}$ and then a second to evaluate s^2 using Equation 32.

4.10 Measurement Distributions

Figure 52 shows two number lines with values marked upon them: one has a small spread and the other a much larger spread. Both have the same mean, however more confidence could be placed in the calculated mean being equal to (or very close) to the true value if the data followed the pattern with the smaller spread: this reinforces the point that the mean does not give us the whole picture when analysing results.

As an alternative to the number line, a set of measurements may be represented by a histogram. Figure 53 shows a histogram for the 10 measurements of the current flowing through a lamp given above. As an increasing number of measurements are taken the intervals (or bin widths) can be made smaller. A curved line (called a distribution curve) can be superimposed on the histogram. The distribution curve passes through the middle of the top of each bar. As the number of measurements tends to infinity the distribution curve

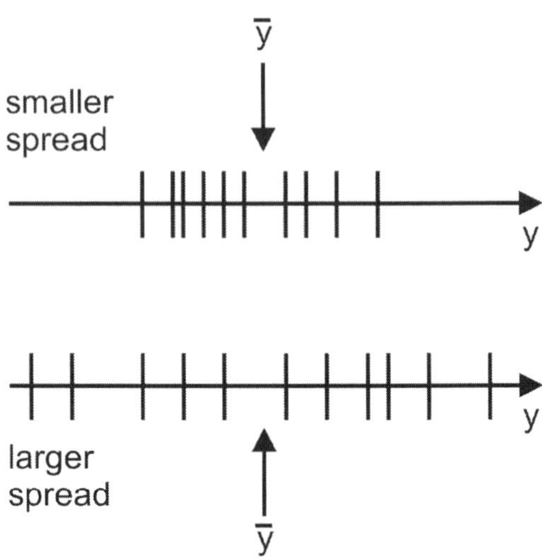

Figure 52: Number lines comparing the spread of data.

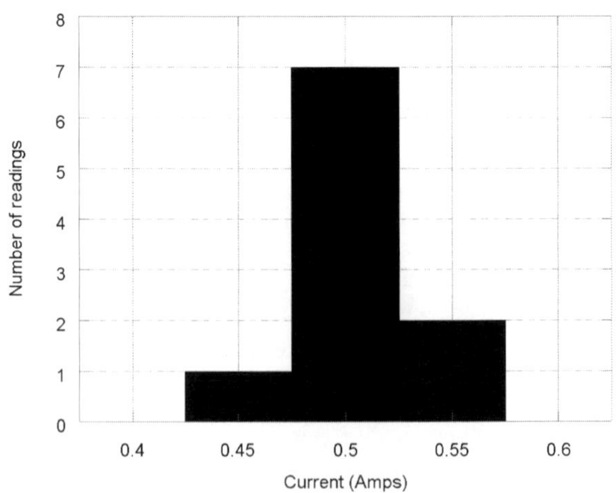

Figure 53: Histogram of the measurements of the current flowing through a bulb given in the text.

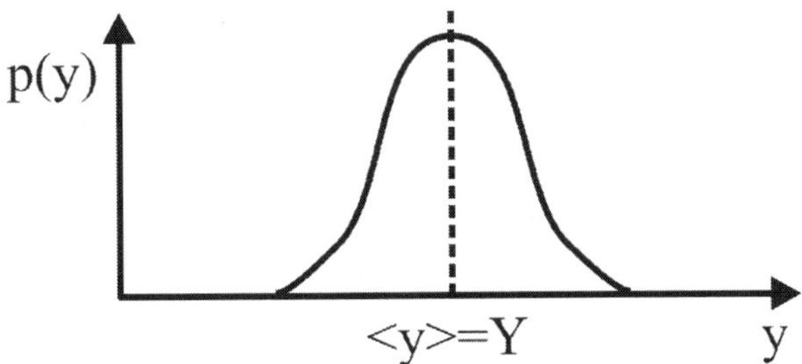

Figure 54: Histogram of the measurements of a quantity when the number of measurements tends to infinity.

becomes smooth and the mean value, $\overline{y}$, tends to the true value, Y: see Figure 54. The distribution function, $p(y)$, shown in Figure 54 gives the fraction of the total measurements in each interval. Thus $p(y)dy$ is the fraction of the total number of measurements which lie in the interval y to $y+dy$. The statement is equivalent to stating that $p(y)dy$ is the probability that when a single measurement is chosen at random it lies in the interval y to $y+dy$. Extending this interval to cover all y values from $-\infty$ to $+\infty$ means the interval must contain our measured value with probability 1, thus:

$$\int_{-\infty}^{+\infty} p(y)dy = 1 \tag{39}$$

and the mean of the distribution, $< y >$, is given by:

$$< y >= \int_{-\infty}^{+\infty} yp(y)dy \tag{40}$$

The distribution shown in Figure 54 is only obtained under 'perfect' circumstances. An infinite number of independent measurements would need to be taken. Clearly, in practice this is impossible and this graph can't be obtained from a finite number of measurements. The mean value, $< y >$, of the distribution shown in Figure 54 is considered to be the true value, Y, of the quantity being measured. The actual set of measurements (for example, the 10 measurements of current from above) is a random sample taken from the distribution, $p(y)$. The sample will have a mean value of $\overline{y}$ (as discussed above) which differs from the true value Y by a random and unknown amount. The point of error analysis is to estimate how much the mean of the sample differs from the true value, that is, to find $|\overline{y} - Y|$ in this notation.

4.11 Standard Error on a Single Measurement

A single measurement is part of a distribution of many single measurements like that shown in Figure 54. The standard error on a single measurement, σ, is the half width of this distribution. When a single measurement is made it will lie somewhere in this distribution. If the spread in Figure 54 is large, then the error in the single measurement will potentially be large and so σ will be large. Conversely, if a the spread in Figure 54 is small, then the error in the single measurement will be small and so σ will be small.

Technically, the standard error on the single measurement is given by:

$$\sigma^2 = \int_{-\infty}^{\infty} (y - Y)^2 p(y) dy \tag{41}$$

In practical terms, using a small set of real data the best estimate of standard error on a single measurement is given by the equation:

$$\sigma = \sqrt{\frac{1}{N-1} \sum_{i=1}^{N} (y_i - \overline{y})^2} = s\sqrt{\frac{N}{N-1}} \tag{42}$$

where there are N different measurements labeled $y_i \ldots y_N$. The standard error on a single measurement does not decrease as the number of measurements increases.

4.12 Standard Error on the Mean

If the errors in the individual measurements are entirely random, then the standard error on the mean value of the N measurements may be found using

$$\sigma_m = \frac{\sigma}{\sqrt{N}} = \frac{s}{\sqrt{N-1}} \tag{43}$$

As the number of measurements taken increases, the standard error on the mean decreases. It follows a square root relationship so increasingly more measurements need to be taken to reduce the error by a small amount.

It is important to understand the difference between the standard error on a single measurement, σ, and the standard error in the mean, σ_m. The standard error on a single measurement is a measure of the distribution of the measured values. This distribution does not become narrower when the number of individual measurements increases: it becomes better defined and the curve becomes smoother. The standard error on the mean does decrease as the number of measurements increases: the mean becomes more accurately known.

Returning to the earlier example in Table 17 the following can now be calculated:

$$\sigma = \sqrt{\frac{6}{5}} \times 9.74 = 10.7 \tag{44}$$

$$\sigma_m = \frac{10.7}{\sqrt{6}} = 4.4 \tag{45}$$

$$\overline{x} = 250 \pm 4 \tag{46}$$

4.13 Estimation of Errors

It often occurs that repeated measurements of a simple quantity like the length of a pendulum gives very little meaningful spread in the results or even the same results each time. In this case calculation of the standard error on a single reading and the standard error in the mean, as described above, is rather meaningless: the accuracy of the measurement is limited by the resolution of the ruler. For example consider using a meter rule with a resolution of 1 cm: that is it has markings every one centimeter. Each time the length is measured the scale is read to the nearest centimeter, so 35.6 would be read as 36. This has introduced a systematic error for this individual reading, but the error in a group of readings is random because it is equally likely to be positive as negative. The error will range from nothing to half a division (0.5 cm), and so the average error will be about a quarter of a division (0.25 cm). The standard error (corresponding to 1 standard deviation if we assume a normal distribution) will be about a third of a division (0.33 cm). Care must be taken to give a reasonable estimate of the error in measurements like this.

4.14 Gaussian Error Distribution

So far estimated values for the the size of errors in measurements have been found. However, nothing has been mentioned about the probability of seeing these errors. Assumptions need to be made that the error is equally likely to be positive as negative and that it follows a Gaussian or Normal distribution.

z	probability
1	0.682689492
2	0.954499736
3	0.997300204
4	0.999936660
5	0.999999427
6	0.999999998

Table 18: Probability that a given measurement is within ‘z’ standard deviations of the mean.

There is no inherent reason why this should be the case, except that in many cases it is found to be true to a good approximation. However, a common discrepancy is that larger errors are more common than would be suggested.

The probability that a certain reading lies between $\pm y_0$ of the true value is

$$P(y) = \frac{1}{\sigma\sqrt{2\pi}} \int_{-y_0}^{y_0} exp\left(-\frac{y^2}{2\sigma^2}\right) dy \tag{47}$$

where σ is the standard error on a single measurement.

Since the integral is difficult to solve analytically, recall the ‘z’ statistic which is looked up in a table of normal distribution values.

$$z = \frac{y - \mu}{\sigma} \tag{48}$$

The ‘z’ value for data being within one standard deviation of the mean is 1, since the difference $y - \mu = \sigma$. This corresponds to a probability of 0.68 or 68 % that a given measurement is within 1 standard deviation of the mean. Table 18 shows the probabilities for higher numbers of standard deviations.

4.15 Combination or Propagation of Errors

If measurements of two or more different parameters are made in an experiment and then used to calculate another quantity the error in the value for this quantity must be calculated. This can be found by using the errors in the measurements of each of the parameters. The simplest case to consider is when there is a linear combination of parameters. More complex situations involve products/quotients or the general case of any function.

4.15.1 Linear

Consider the equation:

$$a = b + c \tag{49}$$

where b and c are the measured parameters. Differentiating gives:

$$\delta a = \delta b + \delta c \tag{50}$$

where δa is the error in a single given measurement of a, similarly for δb and δc. In any given measurement the exact size size of the error is not known: that is, the values of δb and δc are not known. If they were, the errors could simply added or subtracted from the measured value to find the exact values of b, c and then a.

The mean square deviation of a set of measurements from the mean is given by σ^2 giving the mean square error in b as σ_b^2 and the mean square error in c as σ_c^2. These equations apply equally to σ the standard error on a single measurement and σ_m the standard error on the mean of a set of measurements, although for printing clarity the σ version will be used.

If it can be assumed that these measurements of b and c are drawn from a Gaussian distribution, then the maximum error should be infinite since the Gaussian distribution does not come to an abrupt end a certain distance from the mean. A more realistic notion of the maximum error is found by directly adding the errors in b and c giving $\sigma_a = \sigma_b + \sigma_c$. However, this doesn't give the generally accepted rule for combining errors. It may be that the errors instead subtract $\sigma_a = \sigma_b - \sigma_c$ giving a smaller result. Taking a compromise of these two cases involves adding σ_b and σ_c as though they are perpendicular and form two shorter sides on a right angle triangle. Therefore adding in quadrature (i.e. applying Pythagoras' Theorem) gives

$$\sigma_a^2 = \sigma_b^2 + \sigma_c^2 \tag{51}$$

This can be formally proven:

$$\begin{aligned}
\sigma_a^2 &= < [a-\overline{a}]^2 > && (52)\\
&= < \left[(b+c)-(\overline{b}+\overline{c})\right]^2 > && (53)\\
&= < \left[(b-\overline{b})+(c-\overline{c})\right]^2 > && (54)\\
&= < (b-\overline{b})^2 > + < (c-\overline{c})^2 > +2 < (b-\overline{b})(c-\overline{c}) > && (55)\\
&= \sigma_b^2+\sigma_c^2+2 < (b-\overline{b})(c-\overline{c}) > && (56)
\end{aligned}$$

The final term depends on whether the errors in b and c are correlated in some way. If there is no correlation then the size of the error in b has no effect on the size of the error in c and visa versa. This means that the term $(b-\overline{b})(c-\overline{c})$ will average to zero. This reduces the error in σ_a^2 to the form given in Equation 51.

4.15.2 Products and Quotients

Consider the equation:

$$a = b^x c^y \quad (57)$$

where the powers x and y can be any rational or irrational number including positive values, negative values and fractional values. Taking the logarithm of both sides and the differentiating gives

$$\log a = x \log b + y \log c \quad (58)$$

$$\frac{\delta a}{a} = x\frac{\delta b}{b} + y\frac{\delta c}{c} \quad (59)$$

Following the method applied in the linear case above, take the mean square values added in quadrature:

$$\left(\frac{\sigma_a}{a}\right)^2 = x^2\left(\frac{\sigma_b}{b}\right)^2 + y^2\left(\frac{\sigma_c}{c}\right)^2 \quad (60)$$

A proof of this follows the same pattern as the formal proof given in the linear case.

It is interesting to note that in the product case the fractional error in the parameter is found rather than then absolute error as in the linear case.

In the common cases of $a = bc$ or $a = b/c$ Equation 60 simplifies to:

$$\left(\frac{\sigma_a}{a}\right)^2 = \left(\frac{\sigma_b}{b}\right)^2 + \left(\frac{\sigma_c}{c}\right)^2 \tag{61}$$

Again, Equation 60 and 61 are only valid if the errors in b and c are uncorrelated.

It can be demonstrated that the overall error in not the same for correlated and uncorrelated errors by considering the measurement of the diameter, d, of a wire and then using this to calculate the cross sectional area, A, of the wire. Taking the equation for the area as:

$$A = \frac{\pi d^2}{4} \tag{62}$$

Equation 60 gives:

$$\left(\frac{\sigma_A}{A}\right)^2 = 2^2\left(\frac{\sigma_d}{d}\right)^2 = 4\left(\frac{\sigma_d}{d}\right)^2 \tag{63}$$

However if the equation for the area is taken as:

$$A = \frac{\pi d \times d}{4} \tag{64}$$

That is, we associate b and c from Equation 60 separately with d then the errors σ_b and σ_c are certainly correlated (obviously, they are the same). Equation 60 now gives:

$$\left(\frac{\sigma_A}{A}\right)^2 = \left(\frac{\sigma_d}{d}\right)^2 + \left(\frac{\sigma_d}{d}\right)^2 = 2\left(\frac{\sigma_d}{d}\right)^2 \tag{65}$$

This is only half the error found in Equation 63 and does not give a valid estimate of the error in A.

4.15.3 General Functions

Sometimes the analysis of an experiment depends on functions which are more exotic than simple addition, subtraction, division and multiplication such as trigonometric, exponential or logarithmic functions.

Suppose the function f is defined in terms of n different measured quantities x_i:

$$f = f(y_1, y_2, y_3, ..., y_n) \tag{66}$$

The value f_0 should be calculated by setting all the measurements y_i to their measured or mean values. Next, the n different values of f_i can be found in which each parameter y_i in turn is set equal to its measured (or mean) value plus its error. Finally σ_f is calculated using:

$$\sigma_f^2 = \sum (f_i - f_0)^2 \tag{67}$$

This has the effect of combining in quadrature all the individual changes in f caused by moving each variable by its error.

As an example consider the function:

$$f = \tan(x) \tag{68}$$

If x is measured and found to be $89 \pm 0.5°$ then:

$$f_0 = \tan(89) = 57 \tag{69}$$

$$f_1 = \tan(89.5) = 115 \tag{70}$$

Now calculate:

$$\sigma_f^2 = \sum (f_i - f_0)^2 = (57 - 115)^2 = 3364 \tag{71}$$

$$\sigma_f = 58 \tag{72}$$

This gives us an upper limit on the value of f. A lower limit can be found in a similar way using $x = 88.5$. Giving $\sigma_f = 19$. The final result is:

$$f = 57^{+58}_{-19} \tag{73}$$

Notice that a different upper and lower limit has been found and it is entirely appropriate to quote the error like this.

4.15.4 Appropriate Significant Figures in Final Answers and Errors

The number of significant figures given in a final answer should never be more than the least number of significant figures recorded in any of the measurements taken to find the answer.

As an example consider using the equation $R = V/I$ to find the resistance of a wire. The voltmeter reads 3.45 V and the ammeter reads 0.12 A, this means the (electric) potential difference is given to 3 significant figures, however the current is only given to 2 significant figures. The answer should be given to two significant figures, so $R = 29\Omega$.

To consider the error in this value for resistance, let us take the uncertainty in the ammeter reading to be 0.005 A and the uncertainty in the voltmeter reading to be 0.005 V. This indicates a maximum value of $R = 30.04\,\Omega$ and a minimum value of $R = 27.56\,\Omega$. It is not appropriate nor meaningful to state that $R = 29 \pm 1.44\,\Omega$ as the last two digits given make no changes to the value of $29\,\Omega$. Errors and uncertainty ranges should usually be quoted to the same number of decimal places as the given answer. In this case a correct statement would be $R = 29 \pm 1\,\Omega$.

4.16 Plotting Appropriate Graphs

Graphs allow trends to be spotted in data and straight line graphs allow measurements of physical quantities to be made from gradients or intercepts and help with spotting potential systematic errors. It is usually beneficial to plot data on a graph as soon as to is recorded, rather than waiting until the end of the experiment. This means that anomalous data can be spotted quickly and repeated so that large amounts of time are not wasted.

Numbers in standard form can be entered into a computer spreadsheet by replacing, for example, 6.63×10^{-34} with '6.63e-34'.

4.16.1 Drawing Lines of Best Fit and Finding Gradients

A line of best fit may be drawn by hand onto a graph. The aim should be to draw a single smooth line without using repeated short strokes of the pencil. There should be no sharp kinks in the graph unless the data obviously indicates this. The line need not go directly through any of the points and should have approximately the same number of points sitting above the line as below. Sometimes a flexible ruler is useful for drawing curved lines or a ruler with a slit in the middle (available from http://www.bestfitlineruler.com) is useful for straight lines. The gradient of a straight line should be taken using as large a triangle as reasonably practical. For a curved line, the gradient may be found at individual points by taking the gradient of the tangent to the curve at this point.

Computer programmes can be used to draw a line of best fit (sometimes called a trendline) based on a least squares fitting (described later in more detail) to a mathematical function such as $y = mx + c$ or $y = a + bx + cx^2$.

4.16.2 Error Bars on Graphs

Error bars should be added to graphs for all x and y values where possible. They give a measure of the uncertainty in the position of the point on the axes. If a number of repeats are taken for each point then the error bar should extend above the point by a distance corresponding to σ_m and also below the point by a distance corresponding to σ_m. Horizontal error bars are more usually related to the resolution of the of the equipment, rather than measurement error, but never the less should be included on graphs in a similar way.

4.16.3 Straight Line Graphs

A straight line graph follows the form:

$$y = mx + c \tag{74}$$

where m is the gradient and c is the y axis intercept. If an equation can be rearranged into this format and compared with $y = mx + c$ then the values of m and c from the graph can be used to find the matching parameters. For example consider the emf, ϵ, of an chemical cell of internal resistance r where values of the (electric) potential difference V and the current I have been measured:

$$V = \epsilon - Ir \tag{75}$$

$$y = c + mx \tag{76}$$

If the (electric) potential difference is plotted on the y axis and the current on the x axis, the gradient will be $-r$ and the y axis intercept will be ϵ.

Alternatively, consider the time period T of a pendulum of length l:

$$T = 2\pi\sqrt{\frac{l}{g}} \tag{77}$$

Rearranging this into a more convenient format gives:

$$T^2 = \frac{4\pi^2}{g} l \tag{78}$$

$$y = mx \tag{79}$$

If the square of the time period is plotted on the y axis and the length is plotted on the x axis, the gradient of the graph should equal $4\pi^2/g$ and the acceleration due to gravity may be measured. The y axis intercept in this case should be zero. If a significant y axis intercept is found then a systematic error should be suspected in the data.

4.16.4 Log Graphs

Log graphs are usually used because of one, or both, of two main benefits. Using a logarithmic scale on one or both axes of a graph allows data which span a very large range to be plotted in a more meaningful way. Measurements in the range 1 to 10 take up the same proportion of the axis as measurements in the range 10 to 100 or 100 to 1000. If a log scale is used for the x axis data then $\log x$ rather than x must be plotted on the x axis. It is most convenient to plot a log graph using a computer although it may be done on paper using special logarithmic graph paper (which has non-equal spacing between the grid lines).

A second benefit is that by plotting both the x and y values on a log scale a power law in the data can found, since:

$$y = x^n \tag{80}$$

$$\log y = n \log x \tag{81}$$

If $\log y$ is plotted on the y axis and $\log x$ is plotted on the x axis then the data should form a straight line with gradient n.

4.17 Graph Drawing Software

Files of experimental data recorded by computers are usually in a format known as 'tab delimited' text meaning the table of data is arranged with one row per line and there is a tab separating each column of data. Alternative delimiters include spaces, commas or full stops.

Most programmes allow tab delimited text files to be opened. For example in Excel choose 'File' and then 'Open'. Near the 'File name' box, there is a 'Files of type' drop down. Use this to choose the 'Text files' option. Double clicking on the file (which must have the extension *.txt) starts the 'Text Import Wizard' which allows the starting row to be specified and the delimiter to be set. In Kaleidagraph when any text file is opened the 'Text File Input Wizard' begins automatically.

It is sometimes necessary to differentiate a data set. Recall the definition of differentiation:

$$\frac{df(x)}{dx} = \frac{f(x + \delta x) - f(x)}{\delta x} \tag{82}$$

To differentiate a table of data take the difference in y values between the current data point and the next. Divide this by the difference in x values. For instance if x values are in column A of a spreadsheet and y values in column B, the differential is (B2-B1)/(A2-A1) or more generally (B(N+1)-B(N))/(A(N+1)-A(N)).

4.17.1 Excel 2003 and Prior Versions

Graphs may be plotted using a spreadsheet programme such as Microsoft Excel. Care must be taken so that a scatter graph is plotted with x values on the x axis and y values on y axis. It should not be taken for granted that Excel will plot data as you want it.

With Excel 2003 highlight data with the mouse (holding the Ctrl button while using the mouse allows selection of non-contiguous regions). Next, select the 'Insert' menu and then the 'Chart' option. The 'Chat Wizard' is displayed: see Figure 55. In most cases the 'XY scatter' option will need to be selected as this will plot one set of points verses the other. In step 2 of the 'Chart Wizard' the 'Series' tab allows the data plotted on the x and y axis to be changed. This allows the axes to be swapped should Excel plot them the wrong way round. This is shown in Figure 56. Step 3 of the 'Chart wizard' allows a chart title and axis titles to be set and gives a choice which grid lines to display (along with a few other less note-worthy options).

A trend line may be added to the graph by right clicking the mouse on one of the data points and selecting 'Add Trendline': see Figure 57. An appropriate trend line should be selected (often linear) and in the 'Options' tab, the 'Display Equation on Chart' checkbox should be checked.

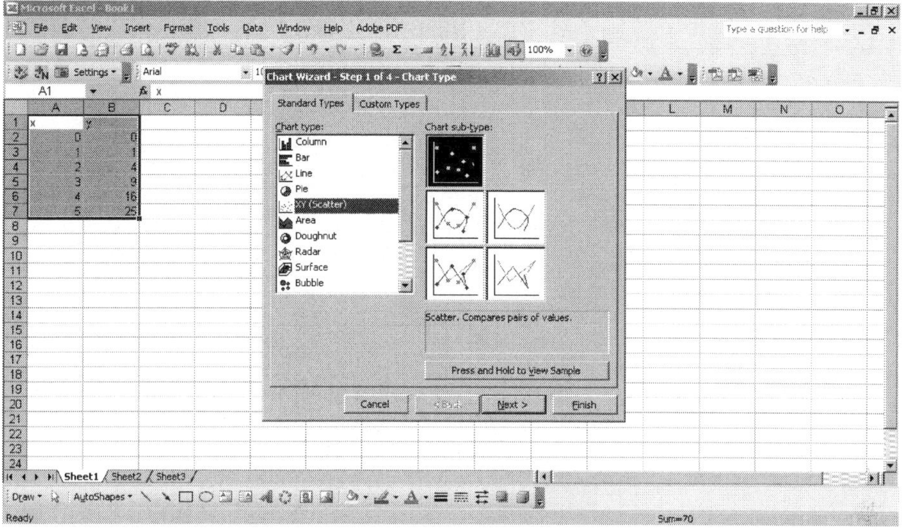

Figure 55: Plotting an 'XY Scatter' graph with Excel 2003.

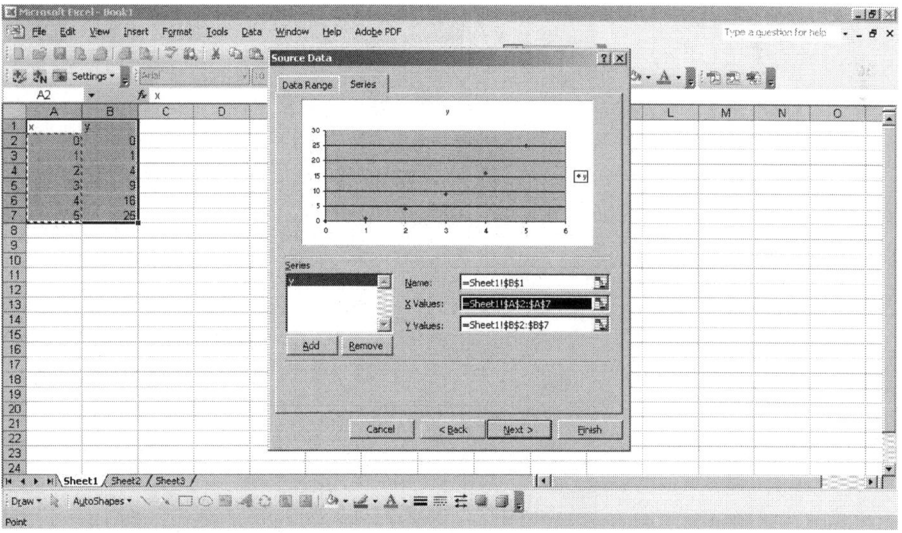

Figure 56: Selecting the data series with Excel 2003.

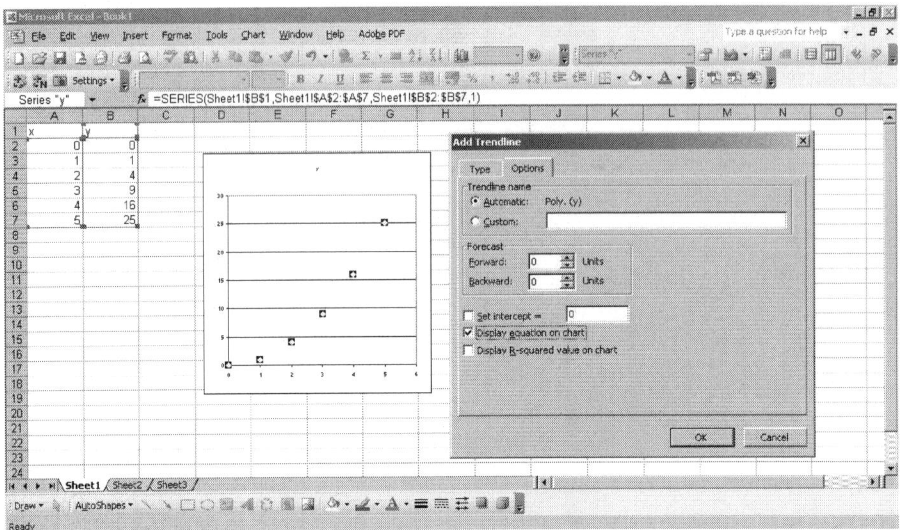

Figure 57: Adding a polynomial trend line of order 2 to the data and ensuring the equation is displayed on the chart.

Error bars can be added to the graph by right clicking on one of the data points and selecting ‘Format Data Series’ from the popup menu. Clicking on the ‘X Error Bars’ tab gives the option to have error bars display in both directions (or just the positive or just the negative directions): see Figure 58. ‘Fixed Value’ error bars can also be chosen (this is generally appropriate where repeated measurements give the same result and the estimated error is based on the resolution), ‘Percentage’ and ‘Custom’ (which allows the selection of a column from the spreadsheet with the same number of rows as the number of x values). Similar options are available for the y axis on the ‘Y Error Bars’ tab.

4.17.2 Excel 2010

With Excel 2010 highlight the data with the mouse (holding the Ctrl button while using the mouse allows selection of non-contiguous regions). Next, select the Insert ribbon and then the ‘Scatter’ option as shown in Figure 59. In most cases the ‘XY Scatter’ option will needed as this will plot one set of points verses the other. This will insert a chart onto the spreadsheet. If Excel has plotted the axes the wrong way round these can swaped by selecting the

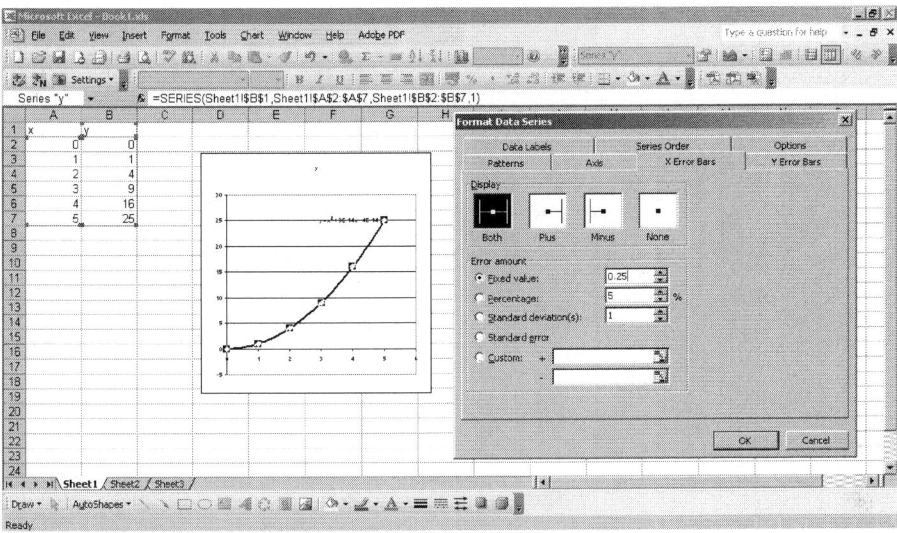

Figure 58: Adding X axis error bars. They have a fixed size of ± 0.25.

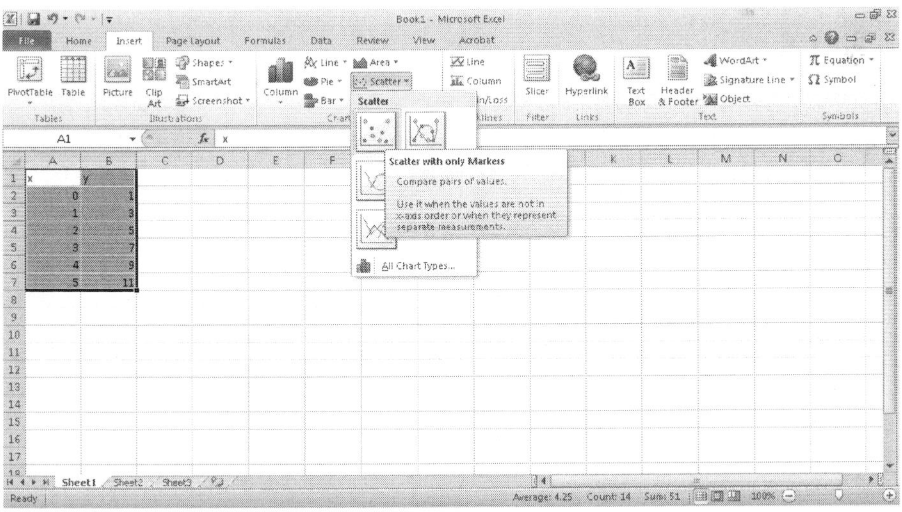

Figure 59: Plotting a graph with Excel 2010

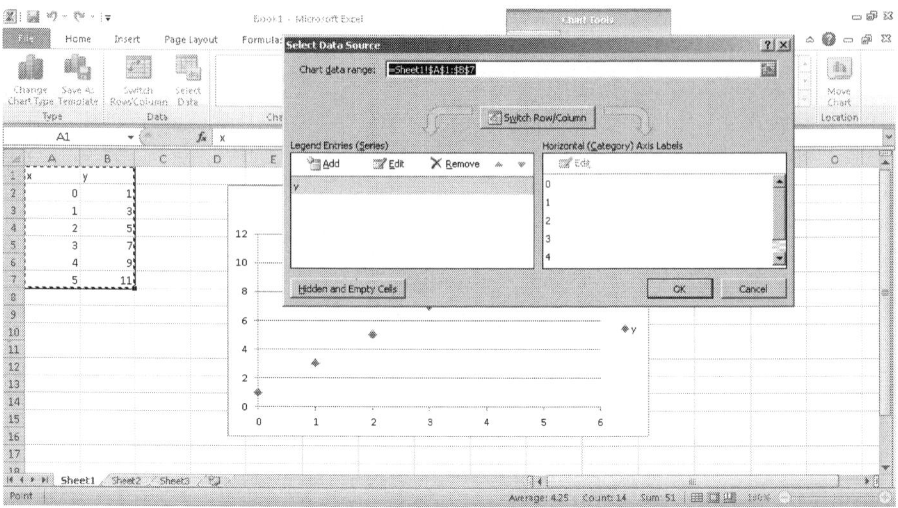

Figure 60: Selecting the data series with Excel 2010.

'Select Data' button from the 'Chart Tools, Design' ribbon. This opens a window which allows the problem to be solved: see Figure 60. To add titles to the x and y axes select the 'Axis Titles' button from the 'Chart Tools, Layout' ribbon as shown in Figure 61. The chart title can be set in the same way.

A trend line may be added to the graph by right clicking the mouse on one of the data points and selecting 'Add Trendline'. An appropriate trend line should be selected (often linear) and in the 'Trendline Options' tab, the 'Display equation on chart' checkbox should be checked: see Figure 62.

Error bars may be added to the graph be choosing the 'Layout' ribbon under 'Chart Tools'. Then choose the small arrow to right of the 'Error Bars' button and select 'More Error Bar Options'. This produces a similar menu as Excel 2003 where there is the choice to have error bars displayed in both directions (or just the positive or just the negative directions): see Figure 63. There is a choice between 'Fixed Value' error bars (this is generally appropriate where repeated measurements give the same result and there is an estimate of the error based on the resolution), 'Percentage' and 'Custom' (which allows the selection of a column from the spreadsheet with the same number of rows as the number of x values). The menu which pops up relates to either vertical or horizontal error bars. To edit the error bars for the other

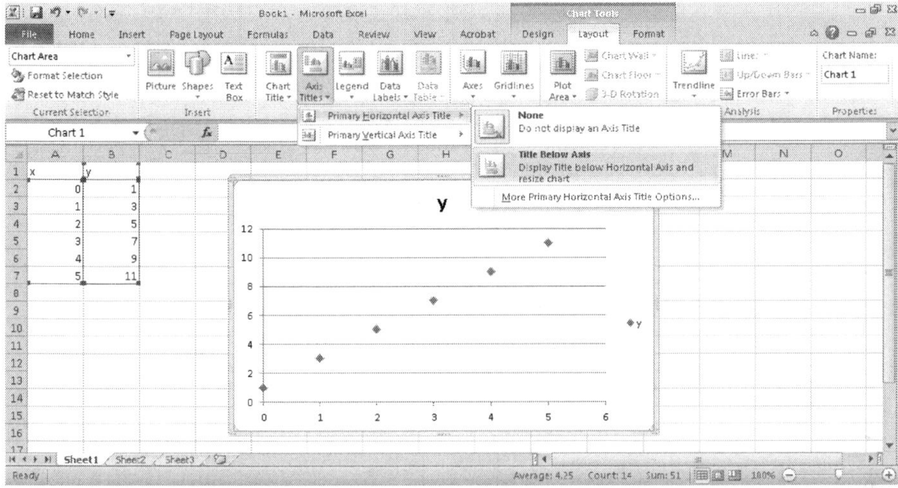

Figure 61: Adding axis titles with Excel 2010.

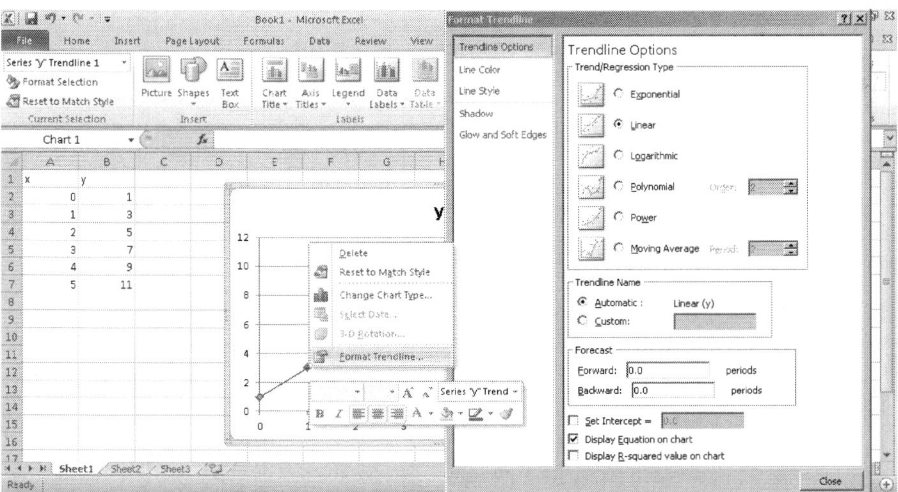

Figure 62: Adding a linear trend line to the data and ensuring the equation is displayed on the chart.

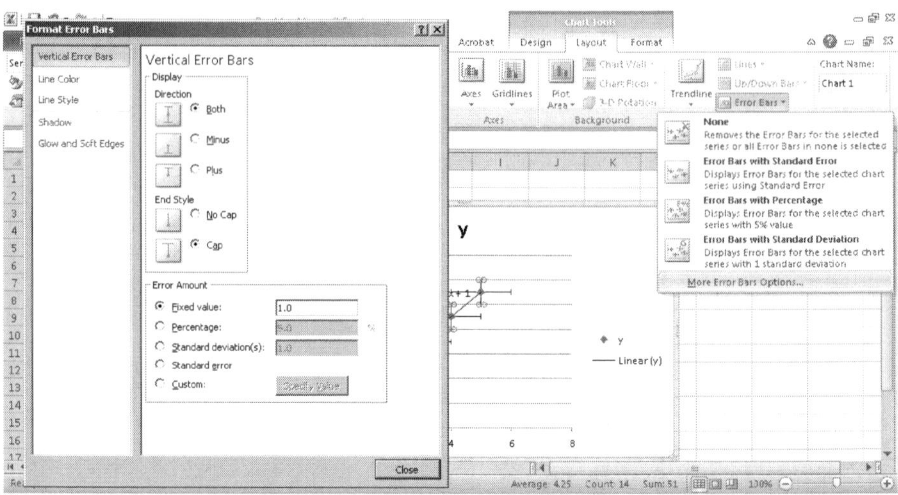

Figure 63: Adding X axis error bars. They have a fixed size of ± 1.

direction, choose the drop down menu in the top left of the 'Layout' ribbon. Either X or Y error bars for each series on the graph can be selected. Now go back to choose the small arrow to right of the 'Error Bars' button and select 'More Error Bar Options': see Figure 64.

4.17.3 Kaleidagraph

In Kaleidagraph, a graph can be plotted by selecting the 'Gallery' menu, followed by 'linear' and then 'scatter': see Figure 65. The next window gives a choice of which variable is plotted on which axes: see Figure 66. Axis and plot titles can be set by double clicking on the default label.

A trend line can be plotted by selecting the 'Curve Fit' menu and then in this case the 'polynomial' option: see Figure 67. By default the fit parameters box is added to the chart. This gives the values of the coefficients m0, m1, and m2 where $y = (m0) + (m1)x + (m2)x^2$. As the data plotted follows a simple $y = x^2$ format, it is no surprise to see that $m0 \approx 0$, $m1 \approx 0$ and $m2 = 1$: see Figure 68.

Error bars can be added by right clicking on the graph and choosing 'Error Bars...' from the popup menu. In the 'Error Bar Variables' window insert a cross in the 'X Err' box to add X error bars. An 'Error Bar Settings' window then pops up in which the type of error bars (Percentage of Value,

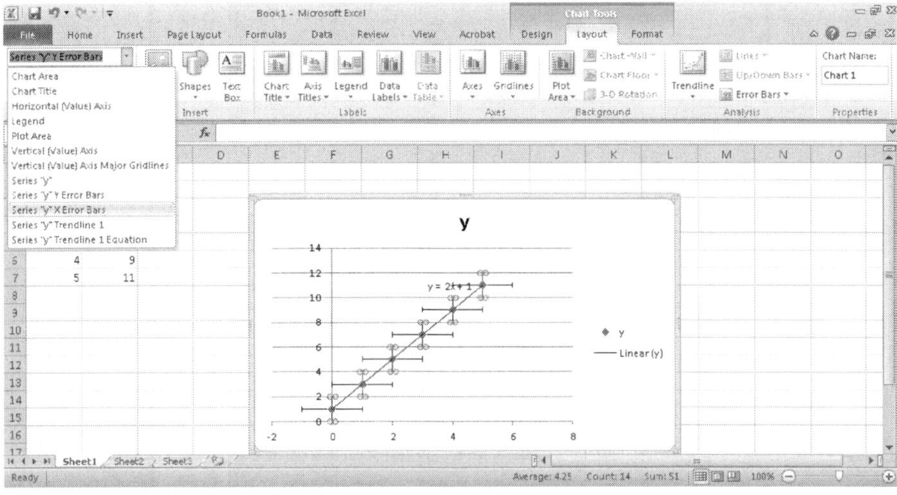

Figure 64: Switching between editing X and Y error bars.

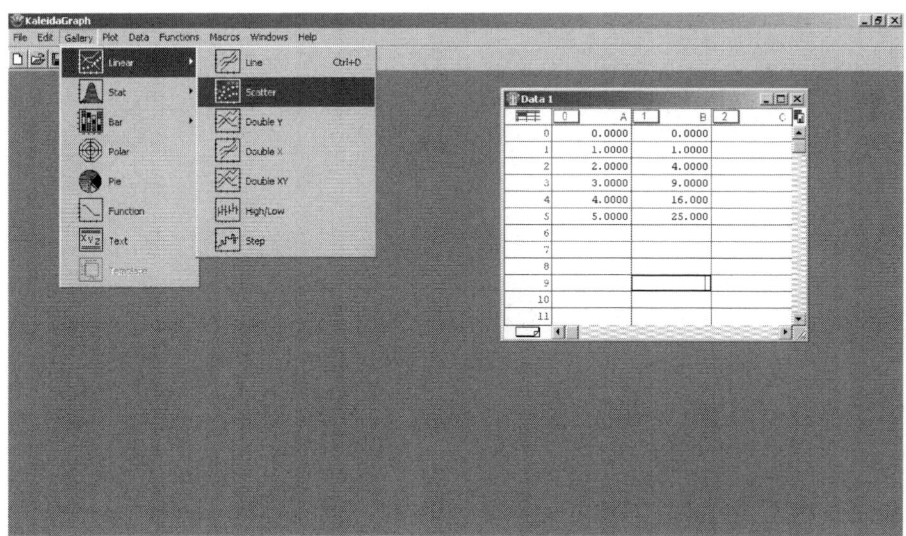

Figure 65: Plotting a graph with Kaleidagraph.

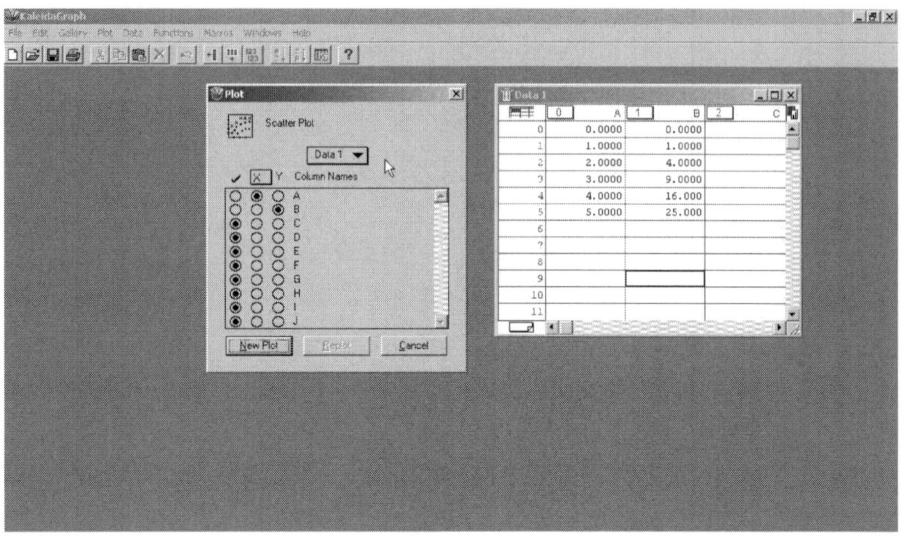

Figure 66: Selecting data for a graph with Kaleidagraph.

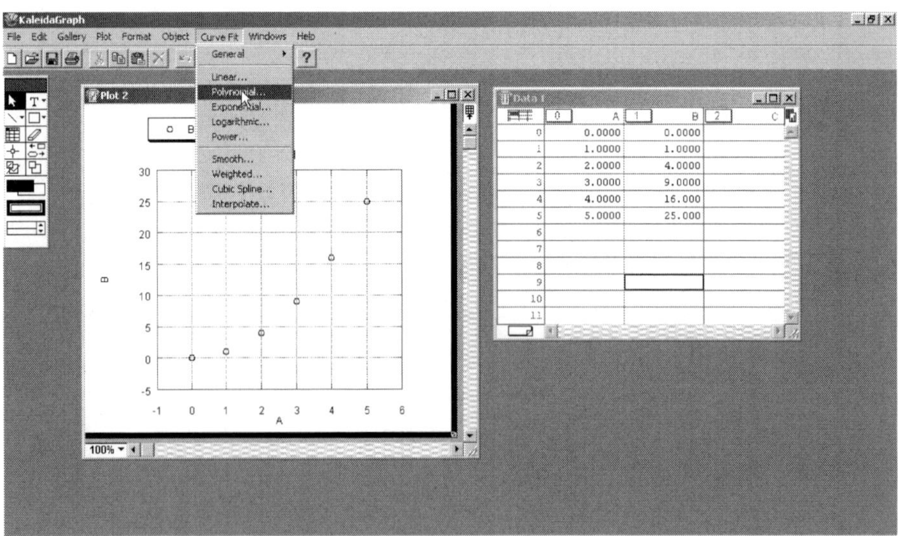

Figure 67: Adding a Trendline with Kaleidagraph

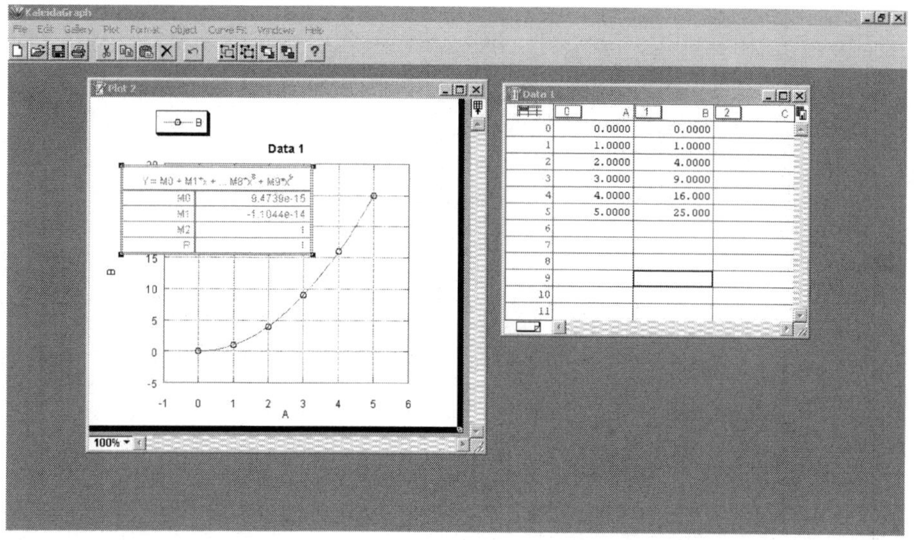

Figure 68: The fit parameters box in Kaleidagraph.

Fixed Value, Standard Deviation, Standard Error or a Data Column) can be chosen. Y error bars can be added by putting a cross in the 'Y Err' box in the 'Error Bar Variables' window. See Figure 69.

A word of caution: graphs and data must be saved as separate files: graphs have the file extension QPC and data tables have the file extension QDA.

4.17.4 Matlab

To draw a graph in Matlab, first load the data points into vectors within the software. A very simple graph can be drawn using the following code which can be entered into the Matlab editor and saved as a *.m* file:

```
i=0:1:100; x=i; y=i.*i; plot(x,y);
```

This produces a graph of $y = x^2$ for x values between 0 and 100 in steps of 1. The x values are stored in the vector x and the y values are stored in the vector y. The simple plot command then produces a graph as shown in Figure 70. Extending this method means more than one line can be plotted on the graph:

```
i=0:1:100; x=i; y=i.*i;

j=0:2:50;
```

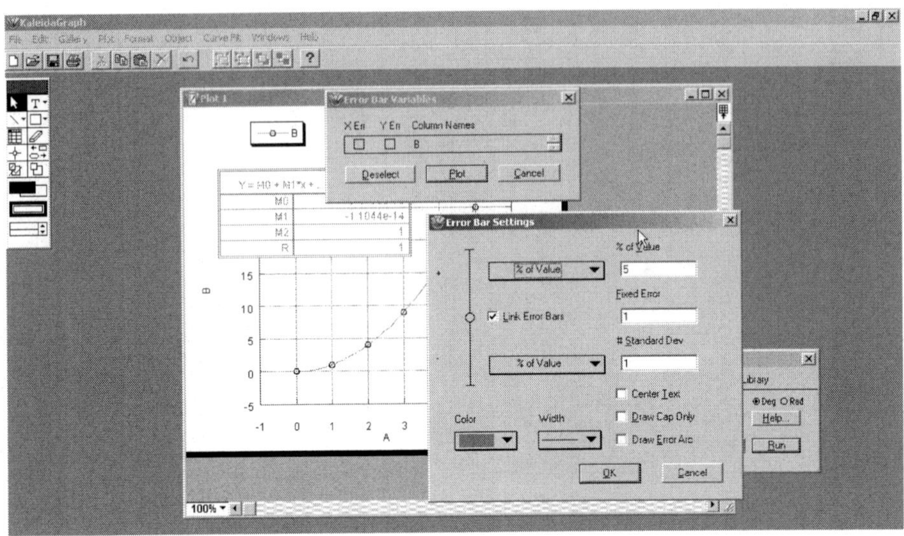

Figure 69: Adding error bars in Kaleidagraph.

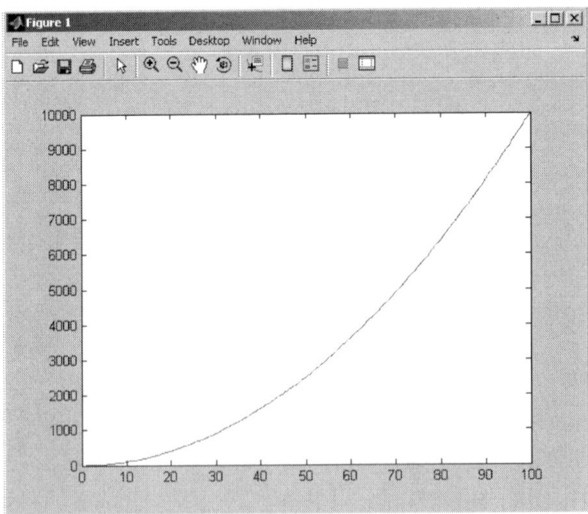

Figure 70: A simple Matlab graph.

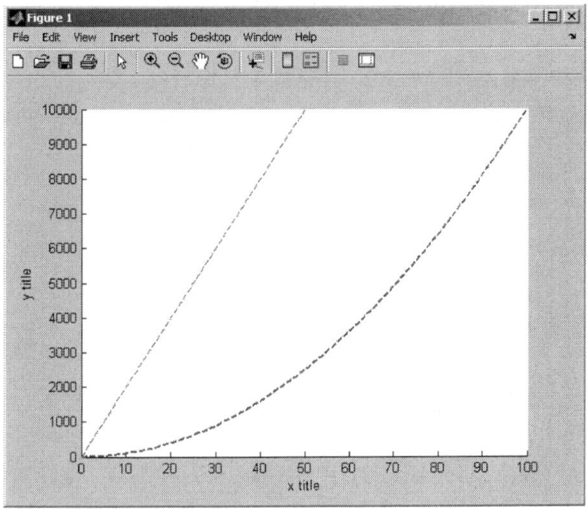

Figure 71: A Matlab graph showing two lines and axes labels.

```
x2=j;
y2=200*j;

hold on plot(x,y);
plot(x2,y2);
hold off
```

The line styles can be changed to dashed and the colours altered to red and blue and axis labels added by typing:

```
i=0:1:100; x=i; y=i.*i;

j=0:2:50;
x2=j;
y2=200*j;

hold on
plot(x,y,'--b','LineWidth',2);
plot(x2,y2,'--r','LineWidth',2);
xlabel('x title')
ylabel('y title')
hold off
```

This produces the graph shown in Figure 71.

In reality data may be saved in text files. If there are five tab delimited text files named file1.dat, file2.dat etc. then they can be loaded and plotted by typing:

```
dataFiles=['file1.dat';'file2.dat';'file3.dat';'file4.dat';'file5.dat'];

hold on
for i = 1:length(dataFiles(:,1))
    data=load(dataFiles(i,:));
    plot(data(:,1),data(:,2));
end
hold off
```

The properties of the graphs such as the line style, line thickness, colours, labels and fonts can be set in the code as illustrated above, but they can also be modified by selecting the 'Tools' menu from the graph window and choosing 'Edit Plot'. Then right click on the graph, the axes or the lines to bring up menus to allow the properties to be modified for the individual graph.

Matlab does not have a built in way of simply adding a best fit line to a graph. One of the least squares method described later in Section 4.18 must be employed. Usually the necessary computer code is already written (often called a numerical recipe) and the function must me called and the results it gives interpreted. A detailed explanation of this is illustrated in Section 4.18.5.

Only vertical errors bars can be added in Matlab. The code needed for this is:

```
i=0:10:100; x=i; y=i.*i; e=0.1.*y; errorbar(x,y,e);
```

which produced the graph shown in Figure 72. A horizontal error bar feature can be added to Matlab by downloading a function called HERRORBAR from the Matlab File Exchange.

4.18 Least Squares Fitting

For a computer the process of obtaining the best fit of experimental data to a line following a particular functional form is basically one of minimising the sum of the square differences between the experimental point and the calculated point for each value of an independent variable. Efficient methods of function minimisation have long been an area of research in computer science resulting in numerous algorithms such as the Steepest Descent method, the Gauss-Newton method, one developed by Levenberg and Marquardt or the Nelder-Mead [16] simplex method.

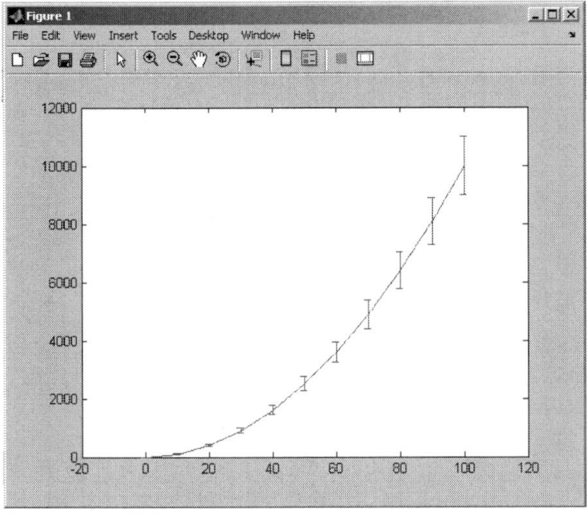

Figure 72: A Matlab graph with vertical error bars.

4.18.1 Regression line of y on x

This method applies to straight line graphs and is often taught as part of the statistics units in school maths courses. The straight line is parameterised as

$$y = a + bx \tag{83}$$

where there are N data points and:

$$a = \overline{y} - b\overline{x} \tag{84}$$

$$b = \frac{S_{xy}}{S_{xx}} = \frac{\sum_{i=0}^{N}(x_i - \overline{x})(y_i - \overline{y})}{\sum_{i=0}^{N}(x_i - \overline{x})^2} \tag{85}$$

This is the method that Microsoft Excel uses to add a linear trendline to a graph. The R^2 value given by Excel provides a measure of the correlation between the data and the trendline. It does not, as is often thought, provide a measure of the goodness of the fit of the line to the data. The R value is

x	y
52.5	58.2
44.4	58.7
55.8	53.3
60.0	47.8
52.3	56.6

Table 19: Example data for a regression line

actually the product moment correlation coefficient where:

$$R = \frac{S_{xy}}{\sqrt{S_{xx}S_{yy}}} = \frac{\sum_{i=0}^{N}(x_i-\overline{x})(y_i-\overline{y})}{\sqrt{\left(\sum_{i=0}^{N}(x_i-\overline{x})^2\right)\left(\sum_{i=0}^{N}(y_i-\overline{y})^2\right)}} \quad (86)$$

$$= \frac{\sum_{i=0}^{N} x_i y_i - \frac{(\sum_{i=0}^{N} x_i)(\sum_{i=0}^{N} y_i)}{N}}{\sqrt{\left(\sum_{i=0}^{N} x_i^2 - \frac{(\sum_{i=0}^{N} x_i)^2}{N}\right)\left(\sum_{i=0}^{N} y_i^2 - \frac{(\sum_{i=0}^{N} y_i)^2}{N}\right)}} \quad (87)$$

A value of $R = 1$ suggests there is perfect correlation between the data and the trendline. A value of $R =$-1 suggests that there is perfect negative correlation (as x increases, y decreases). A value of $R = 0$ suggests there is no correlation. It should also be noted that correlation in no way suggests that there is causation. For example there may be a correlation between the number of cars driving across a road junction in Cape Town and the sales at a well known London department store, but there is no suggestion that drivers in Cape Town are making their way to London!

Consider plotting the data points (0,4.9), (5,5.1) and (10,4.9) which are scattered about the horizontal line y = 5 and fitting a linear trendline, observe that $R^2 \rightarrow 0$. This is despite there being a near perfect fit between the trendline and the data. This confirms that the R^2 value is not a measure of the goodness of the fit, but one of correlation.

Take the example of the data shown in Table 19. The result of the calculations is given in Table 20. Substituting these into the Equation 86 gives $R =$-0.868 so $R^2 = 0.753$. Equations 84 and 85 can then be used to find values for $a = 91.1$ and $b =$-0.682 and finally the equation of the line of best fit:

$$y = 91.1 - 0.682x \quad (88)$$

N	5
Σx	265.0
Σy	274.6
Σx^2	14176.54
Σy^2	15162.22
Σxy	14464.10

Table 20: Example calculations for a regression line

This is precisely what Excel finds when it plots a trendline for this data as shown in Figure 73.

4.18.2 Steepest Descent

This is a relatively simple method based on following the direction in which the gradient is steepest. As an example take a function such as $f(x, y) = 2x^4 - x^3 + y^2 + 6$. Find the vector $\nabla f(x, y)$, this points uphill, so take $-\nabla f(x, y)$ which must point downhill. Recall that

$$\nabla = \Sigma_{i=1}^{N} \hat{e}_i \frac{\partial}{\partial x_i} \tag{89}$$

which is

$$\nabla = \hat{x}\frac{\partial}{\partial x} + \hat{y}\frac{\partial}{\partial y} + \hat{z}\frac{\partial}{\partial z} \tag{90}$$

is three dimensional cartesian coordinates. So in this example

$$\nabla f(x, y) = \left[8x^3 - 3x^2\right] \hat{x} + [2y]\, \hat{y} \tag{91}$$

Taking the starting point as (a, b), the point (a', b') must be found such that $a' = a - \gamma \nabla f(a, b)$ and $b' = b - \gamma \nabla f(a, b)$. γ is the step size, if this is small enough then $f(a, b) > f(a', b')$ and so by finding $f(a', b')$ a smaller value for the function $f(x, y)$ has been found. This procedure can be reproduced iteratively until, on a given iteration $f(a, b)$ and $f(a', b')$ differ by less than a set tolerance. Often the function $f(a, b)$ would be the sum of the square differences between a set of experimental data points and a set of data points derived from a equation of a line of best fit. The steepest descent method is simple, easy to apply computationally and each step is very fast. It is also quite efficient at finding the minimum point if just one exists. If more than

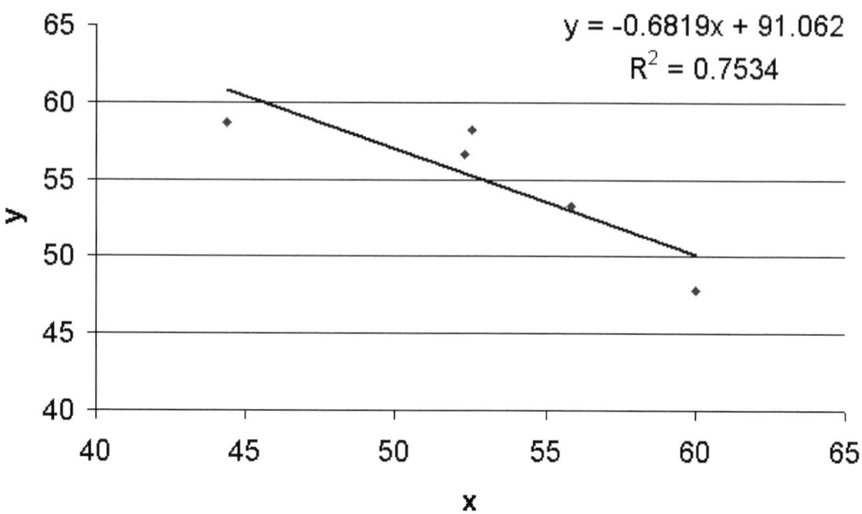

Figure 73: An Excel trendline, showing the equation of the straightline and the R^2 value are given by Equations 84, 85 and 86.

one minimum point exists, or if there is a local minimum, this method may struggle or reach different answers depending upon the starting point. The downside of this algorithm is that the progress towards the minimum gets slower and slower and often a very large number of steps are required. Also, this only works where the function is differentiable. If this differentiation needs to applied numerically, the algorithm can become quite slow. In the case of fitting a function to experimental data the function $f(x, y)$ would be the sum of the square differences between the experimental data and the data generated by the fitting function.

4.18.3 Gauss-Newton

Assume there are M variables in the function $r(\mathbf{b})$ so that the initial parameter vector is $\mathbf{b}_0 = (b_1, ..., b_M)$. The function $r(\mathbf{b})$ is the difference between the experimental value and the fit value given by the fitting function.

$$r_i = y_i - f(x_i, \mathbf{b}) \tag{92}$$

If there are N experimental data points there will be N different functions $r(\mathbf{b})$. As long as $N \geq M$ the Gauss-Newton method will iteratively find the minimum of the sum of the square differences:

$$S(\mathbf{b}) = \sum_{i=1}^{N} r_i^2(\mathbf{b}) \tag{93}$$

To begin the calculation define:

$$J(\mathbf{b}) = \begin{pmatrix} \frac{\partial r_1}{\partial b_1}(\mathbf{b}) & \cdots & \frac{\partial r_1}{\partial b_M}(\mathbf{b}) \\ \vdots & \ddots & \vdots \\ \frac{\partial r_N}{\partial b_1}(\mathbf{b}) & \cdots & \frac{\partial r_N}{\partial b_M}(\mathbf{b}) \end{pmatrix} \tag{94}$$

So the iterative formula used in the Gauss-Newton method is:

$$\mathbf{b}_{k+1} = \mathbf{b}_k - \left(J(\mathbf{b}_k)^T J(\mathbf{b}_k)\right)^{-1} J(\mathbf{b}_k)^T r(\mathbf{b}_k) \tag{95}$$

where $\mathbf{b}_k$ is the k-th modified version of the parameter vector.

Let's try an example. Suppose there is some experimental data:

$$(x_1, y_1) = (1, 2.1663) \tag{96}$$
$$(x_2, y_2) = (2, 2.6726) \tag{97}$$
$$(x_3, y_3) = (5, 5.0180) \tag{98}$$
$$(x_4, y_4) = (7, 7.6373) \tag{99}$$
$$(x_5, y_5) = (9, 11.6236) \tag{100}$$

and it is to be fitted to the equation:

$$y_i = b_1 e^{b_2 x_i} \tag{101}$$

Firstly set up the functions $r(\mathbf{b})$ where $\mathbf{b} = (b_1, b_2)$:

$$r_1(b_1, b_2) = 2.1663 - b_1 e^{b_2} \tag{102}$$
$$r_2(b_1, b_2) = 2.6726 - b_1 e^{2b_2} \tag{103}$$
$$r_3(b_1, b_2) = 5.0180 - b_1 e^{5b_2} \tag{104}$$
$$r_4(b_1, b_2) = 7.6373 - b_1 e^{7b_2} \tag{105}$$
$$r_5(b_1, b_2) = 11.6236 - b_1 e^{9b_2} \tag{106}$$

Next calculate:

$$J(\mathbf{b}) = J(b_1, b_2) = \begin{pmatrix} -e^{b_2} & -b_1 e^{b_2} \\ -e^{2b_2} & -2b_1 e^{2b_2} \\ -e^{5b_2} & -5b_1 e^{5b_2} \\ -e^{7b_2} & -7b_1 e^{7b_2} \\ -e^{9b_2} & -9b_1 e^{9b_2} \end{pmatrix} \tag{107}$$

Taking a starting 'guess' of $\mathbf{b}_0 = (b_1, b_2)_0 = (1.8, 0.2)$:

$$r_1(\mathbf{b}_0) = -0.0322 \tag{108}$$
$$r_2(\mathbf{b}_0) = -0.0127 \tag{109}$$
$$r_3(\mathbf{b}_0) = 0.1251 \tag{110}$$
$$r_4(\mathbf{b}_0) = 0.3379 \tag{111}$$
$$r_5(\mathbf{b}_0) = 0.7342 \tag{112}$$

and then:

$$J(\mathbf{b}_0) = \begin{pmatrix} -1.2214 & -2.1985 \\ -1.4918 & -5.3706 \\ -2.7183 & -24.4645 \\ -4.0552 & -51.0955 \\ -6.0497 & -98.0043 \end{pmatrix} \tag{113}$$

So:

$$\mathbf{b}_1 = \mathbf{b}_0 - \left(J(\mathbf{b}_0)^T J(\mathbf{b}_0)\right)^{-1} J(\mathbf{b}_0)^T r(\mathbf{b}_0) \tag{114}$$
$$= \mathbf{b}_0 - (0.0467, -0.0104) \tag{115}$$
$$= (1.8467, 0.1896) \tag{116}$$

The process is then repeated to find $\mathbf{b}_2$, $\mathbf{b}_3 \ldots \mathbf{b}_z$ where there are z iterations. If z is large enough it will eventually converge to $(1.800, 0.200)$.

The Gauss-Newton algorithm may converge slowly or not at all if the initial guess, $\mathbf{b}_0$, is far from the minimum and it can only be used to minimise the sum of the squares differences of a function. However, it can converge very quickly and no second derivatives need to be calculated.

4.18.4 Levenberg-Marquardt Method

The Levenberg-Marquardt algorithm is more robust than the Steepest Descent and the Gauss-Newton methods as it can find solutions even if the starting 'guess' is a long way from the solution. If there is only one minimum, a solution (or convergence) is guaranteed no matter what the starting 'guess'. It uses a damping factor, λ, which is adjusted each iteration. If the reduction in the sum of the square differences, $S(\mathbf{b})$ is large a smaller value for λ may be used on the next iteration, bringing the algorithm closer to the GaussNewton method. But, if there is little change is the value of $S(\mathbf{b})$ then λ can be increased on the next iteration giving similar behaviour to the Steepest Descent method following the direction of the gradient.

Similar to the Guass-Newton method:

$$S(\mathbf{b}) = \sum_{i=1}^{N} r_i^2(\mathbf{b}) \tag{117}$$

must be minimised where r_i is the difference between the experimental value and the fit value given by the fitting function. If there are N experimental data points there will be N different functions $r(\mathbf{b})$. In each iteration step, the parameter vector $\mathbf{b}$ is replaced by an improved estimate $\mathbf{b} + \boldsymbol{\delta}$. To find the vector $\boldsymbol{\delta}$, take the first order terms from a Taylor expansion:

$$f(x_i, \mathbf{b} + \delta) \approx f(x_i, \mathbf{b}) + J_i\boldsymbol{\delta} \tag{118}$$

where J_i is the same as in the Gauss-Newton method:

$$J(\mathbf{b}) = \begin{pmatrix} \frac{\partial r_1}{\partial b_1}(\mathbf{b}) & \cdots & \frac{\partial r_1}{\partial b_M}(\mathbf{b}) \\ \vdots & \ddots & \vdots \\ \frac{\partial r_N}{\partial b_1}(\mathbf{b}) & \cdots & \frac{\partial r_N}{\partial b_M}(\mathbf{b}) \end{pmatrix} \tag{119}$$

So:

$$S(\mathbf{b} + \boldsymbol{\delta}) \approx \sum_{i=1}^{N} (y_i - f(x_i, \mathbf{b}) + J_i\boldsymbol{\delta})^2 \tag{120}$$

when written in terms of vectors:

$$S(\mathbf{b} + \boldsymbol{\delta}) \approx |\mathbf{y} - f(\mathbf{x}, \mathbf{b}) - J\boldsymbol{\delta}|^2 \tag{121}$$

Consider that, at the minimum of the sum of the square differences $S(\mathbf{b})$, the gradient of the sum of the square differences with respect to $\boldsymbol{\delta}$ is zero. If this was not the case, then it would not be at a minimum. Therefore:

$$J^T J\boldsymbol{\delta} = J^T\left[\mathbf{y} - f(\mathbf{x}, \mathbf{b})\right] \tag{122}$$

which can be solved for $\boldsymbol{\delta}$ by simply taking the inverse of $J^T J$ and applying it to both sides.

Levenberg modified this slightly to introduce the damping term mentioned earlier:

$$(J^T J + \lambda I)\boldsymbol{\delta} = J^T\left[\mathbf{y} - f(\mathbf{x}, \mathbf{b})\right] \tag{123}$$

where I is the identity matrix. If λ is large, then $J^T J + \lambda I \approx \lambda I$ and the calculation of the inverse of $J^T J + \lambda I$ is not used.

Marquardt's modification was to replace the identity matrix with a matrix consisting of the diagonal components of $J^T J$ which has the effect of scaling each component of $\boldsymbol{\delta}$ by the curvature so that bigger steps are taken when the gradient is smaller avoiding slow convergence. Thus the Levenberg-Marquardt algorithm is:

$$(J^T J + \lambda \text{diag}(J^T J))\boldsymbol{\delta} = J^T\left[\mathbf{y} - f(\mathbf{x}, \mathbf{b})\right] \tag{124}$$

The algorithm proceeds iteratively until any one of a number of conditions are met. These are often related to the magnitude of $\boldsymbol{\delta}$ being lower than a certain threshold, the magnitude of $[\mathbf{y} - f(\mathbf{x}, \mathbf{b})]$ becoming smaller than a certain threshold, a maximum number of iterations being completed or the value of $J^T\left[\mathbf{y} - f(\mathbf{x}, \mathbf{b})\right]$ being lower than a certain threshold.

4.18.5 Nelder-Mead Method

As a demonstration consider a function with two parameters so the simplex is a triangle. The function to be minimised, S, is first evaluated at each of the three vertices of the triangle and the vertices ordered B(est), G(ood), W(orst) with increasing values of S. From W to B and W to G the value of the function decreases so it's feasible that S is smaller at values away from W so the triangle is reflected about the line joining B and G giving a new vertex R. The point M is defined as the point halfway between B and G (see Figure 74a).

Consider the case where $S(R) < S(G)$. If $S(B) < S(R)$ then W is replaced with R, otherwise the simplex is extended. The line joining M and R is

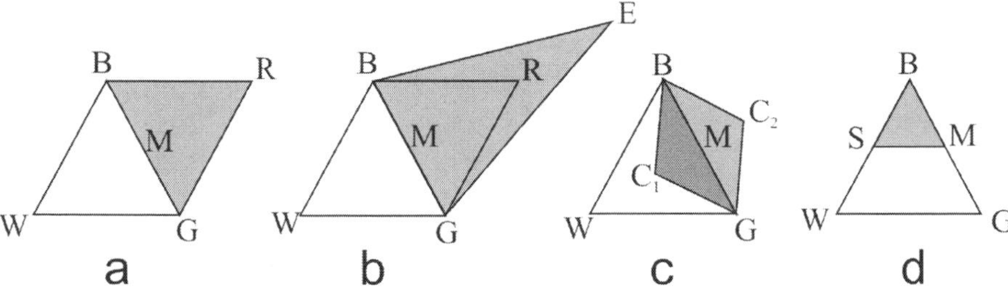

Figure 74: The different triangles involved in a two parameter version of the Nelder-Mead function minimisation algorithm.

extended to point E. Now, if $S(E) < S(B)$ then W is replaced with E, otherwise W is replaced with R (see Figure 74b). The process now repeats by calculating a new R point until the function falls below a pre-set tolerance or is evaluated a certain number of times.

The alternative situation is $S(R) > S(G)$. If $S(R) < S(W)$ then W is replaced with R, otherwise the simplex is contracted. The points C_1 and C_2 are calculated (see Figure 74c) and the point with the smaller function value is called C. If $S(C) < S(W)$ then W is replaced with C, otherwise the point S is calculated as halfway between B and W and the simplex is shrunk to BMS (see Figure 74d). This process repeats until a predefined convergence criterion is reached such as S becoming less than a certain value, failing to change between successive iterations or a maximum number of iterations is reached. Matlab implements this method using its *fminsearch* routine in the optimisation toolbox. The Matlab function *fminsearch* can be modified to calculate and return the covariance matrix i.e. the errors on the fit parameters using the method outlined in the appendix of [16]. The declaration at the top is modified to include an additional parameter N which is passed in the function call which gives the total number of data points and also to return the error vector.

```
function [x,errorvector,fval,exitflag,output] =
fminsearchmf(funfcn,x,N,options,varargin)
```

This additional code was added after line 382 in the code for *fminsearch* and is based on the appendix of [16].

```
    %calculate covariance matrix
    %qmatrix
    temp=size(v);
```

```
numberofparams=temp(1,1);
for i=1:numberofparams
    qmatrix(i,:)=v(i,:)-v(i,1);
end
qmatrix(:,1)=[];
%bmatrix
for i=1:numberofparams
    for j=1:numberofparams
        if i==j
            yi=fv(1,i+1);
            y0=fv(1,1);
            p0i=(v(:,1)+v(:,i+1))./2;
            y0i=funfcn(p0i,varargin{:});
            bmatrix(i,i)=2*(yi+y0-2*y0i);
            a(i)=2*y0i-(yi+3*y0)/2;
        else
            pij=(v(:,i+1)+v(:,j+1))./2;
            yij=funfcn(pij,varargin{:});
            yi=fv(1,i+1);
            y0=fv(1,1);
            p0i=(v(:,1)+v(:,i+1))./2;
            p0j=(v(:,1)+v(:,j+1))./2;
            y0i=funfcn(p0i,varargin{:});
            y0j=funfcn(p0j,varargin{:});
            bmatrix(i,j)=2*(yij+y0-y0i-y0j);
            a(i)=2*y0i-(yi+3*y0)/2;
        end
    end
end
covarmatrix=qmatrix*inv(bmatrix)*qmatrix';
ymin=y0-a*inv(bmatrix)*a';
normalisefactor=2*ymin/(N-numberofparams);
covarmatrix=covarmatrix.*normalisefactor;
%get standard deviation
for i=1:numberofparams
    errorvector(i)=sqrt(covarmatrix(i,i));
end
%end covariance matrix calculation
```

A working skeletal example of this procedure in action is shown below so that it may be copied and modified as necessary. In order for this to work in its current form it is necessary to complete the modification to the file *fminsearch.m* described above.

The 11 data points in Table 21 are generated using the equation $y = ax^2 + b$ where $a = 46$ and $b = 3$: These are saved as a tab delimited text file

x	y
0	0
1	49
2	187
3	417
4	739
5	1153
6	1659
7	2257
8	2947
9	3729
10	4603

Table 21: Example data for Nelder-Mead method.

data.txt in the same directory as the Matlab script quoted below.

```
function fitprm = example()

options=optimset('Display','iter'); %set options for fminsearch
x=0:1:10; %set x points at which to simulate data
Data = load('data.txt'); %load data file
expData = interp1(Data(:,1),Data(:,2),x,'spline');
%interpolate data to values at simulation points given by x
params(1)=40; %starting parameter
params(2)=5; %starting parameter

numberofdatapoints=length(x);
disp(['initialguess='num2str(params)])
estimates=fminsearch(@fit,params,numberofdatapoints,options,x,expData);
disp('Estimates')
disp(num2str(estimates))
fitprm = estimates;

function differenceVector = fit(params,x,expData)
disp(params);
fitData =params(1)*x.*x+params(2);
differenceVector=sum((fitData-expData).^2);
```

When this script is run, it performs the Nelder-Mead simplex method. As the script is in operation the procedures it follows are detailed in the Matlab command window. Once it has converged on a solution it outputs the answers $a = 46.0089$ and $b = 2.4174$ and the uncertainty in the results $\sigma_a = 0.0117$

and $\sigma_b = 0.5595$. It is clear the parameter a has been accurately identified and has a very low uncertainty and the parameter b has been slightly less well identified however the uncertainty is correspondingly larger.

4.18.6 Distribution Testing

Distribution testing is a useful way to see if measured data supports a certain conclusion. Suppose that a test is carried out to see if the volume of a gas remains constant as it is heated. It would be more rigorous to test whether the graph of volume against temperature is consistent with the volume being constant rather than just testing whether a straight line gives a fit where the gradient is zero. For example the volume may have a sinusoidal dependence on temperature (for some reason!). A straight line fitted to this data would certainly give a gradient close to zero, but the volume is certainly not constant.

Distributions can be tested with the χ^2 parameter:

$$\chi^2 = \sum_i^N \left(\frac{y_{calculated} - y_i}{\sigma_i} \right)^2 \tag{125}$$

where there are N data points, y_i is the experimental value, $y_{calculated}$ is the theoretical y value to compare with and σ_i is the error in y_i. Without some idea of the magnitude of σ_i it is impossible to test if y_i and $y_{calculated}$ are consistent.

Calculate the number of degrees of freedom using:

$$\nu = N - p \tag{126}$$

where p is the number of parameters allowed to vary in fitting the data. A limit can be set, typically, 5%=0.05 as an acceptance criterion. Using the values of the acceptance criterion and ν a value for χ^2 can be found from tables.

For example, with $\nu = 1$, if the χ^2 value found from tables is 3.8 this suggests that there is only a 5% probability that the experimental value for χ^2 would be bigger than 3.8. If the χ^2 value calculated for the experiment is less than 3.8, then it is reasonable to state that the theoretical and experimental distributions are consistent at the 95% level. If the χ^2 value calculated for the experiment is more than 3.8, then it is reasonable to state that the theoretical and experimental distributions are not consistent at the 95% level.

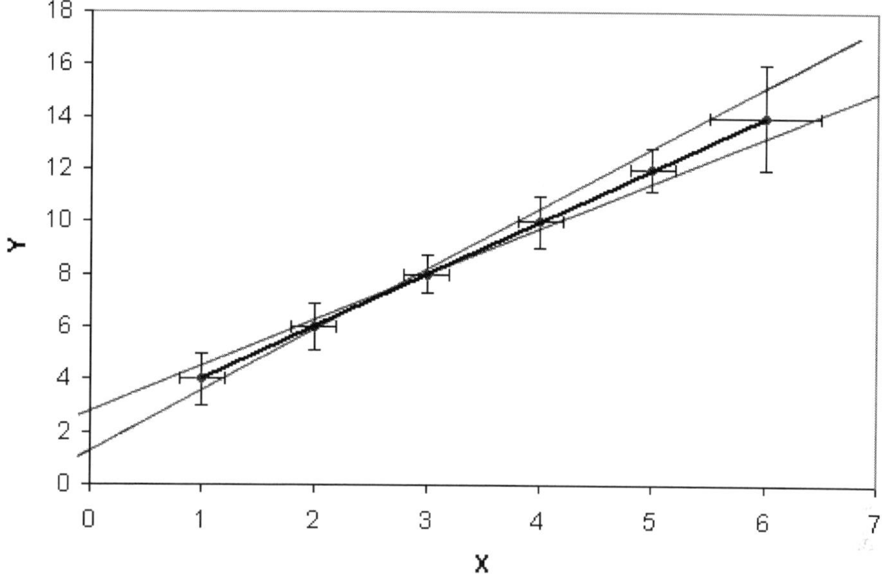

Figure 75: A graph showing the maximum and minimum gradients consistent with the error bars.

4.19 Uncertainty in Gradients and Intercepts of Graphs

Errors in gradients and intercepts of graphs can be found following a simple method of drawing the lines of best fit with the maximum and minimum possible gradients consistent with the error bars plotted around the points. Figure 75 shows a set of example data plotted with x and y error bars. The points follow the line $y = 2.0x + 2.0$ and so the gradient is 2.0 and the y intercept it 2.0. The maximum gradient can be found by drawing a line (by hand or manually using the drawing tool in Excel) through the top/left of the error bars on the point (6,14) and the bottom/right of the error bars on the point (1,4). The line therefore goes through the points (5.5,14) and (1.2,4): it has gradient 2.3 and y intercept of 1.2. The minimum gradient can be found by drawing a line through the bottom/right of the error bars on the point (6,14) and the top/left of the error bars on the point (1,4). The line therefore goes through the points (6.5,14) and (0.8,4): it has gradient 1.8 and y intercept of 2.6. The gradient should be given as $2.0^{+0.3}_{-0.2}$ or perhaps as 2.0 ± 0.3 and the intercept should be given as $2.0^{+0.6}_{-0.8}$ or perhaps as 2.0 ± 0.8.

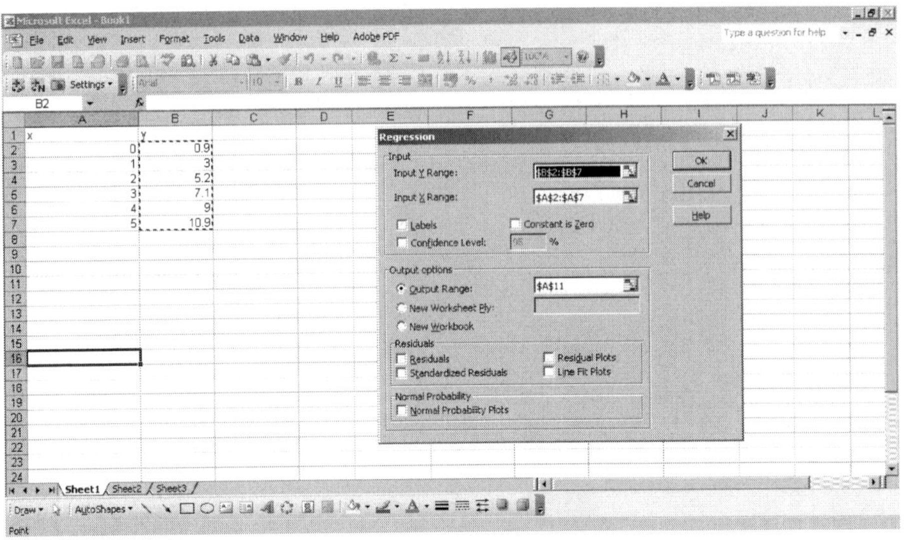

Figure 76: The Microsoft Excel 2003 Regression Window.

More formally, for the case of a straight line the uncertainty in the gradient and intercept can be given (with no derivation here) as:

$$(\Delta\text{gradient})^2 \approx \frac{1}{\Sigma_{i=1}N(x_i - \overline{x})^2} \frac{\Sigma_{i=1}Nr_i^2}{N-2} \tag{127}$$

and

$$(\Delta\text{intercept})^2 \approx \left(\frac{1}{N} + \frac{\overline{x}^2}{\Sigma_{i=1}N(x_i - \overline{x})^2}\right) \frac{\Sigma_{i=1}Nr_i^2}{N-2} \tag{128}$$

where there are N data points and r_i is the difference in y values between the experimental data and the straight line for each x value.

The Microsoft Excel Regression package provides a useful way to estimate errors in the gradient and intercept of straight line graphs. This package is not always loaded into Excel. In Excel 2003 click on 'Tools' and then choose 'Ad Ins'. Then select the 'Analysis ToolPak' checkbox. The option to select 'Data Analysis' is available from the 'Tools' menu. Find and select the 'Regression' option. This brings up the 'Regression' window shown in Figure 76. Once the 'Input X Range' and the 'Input Y Range' have been set and the 'Output Range' selected, the calculations begin. The output looks similar to Figure 77. In Figure 77 the intercept and gradient are given in cells B27 and B28. Cells C27 and C28 give the standard error on these values.

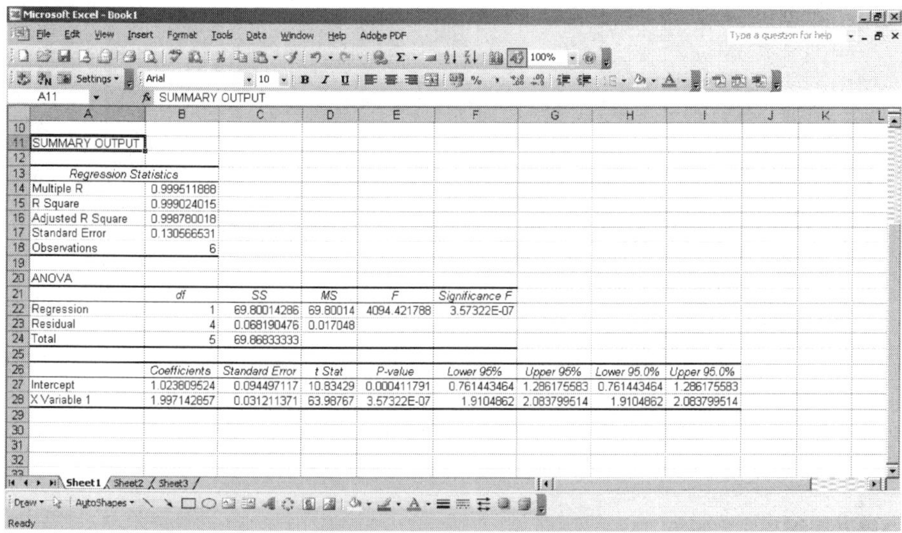

Figure 77: The Microsoft Excel 2003 Regression package output.

In Excel 2010 click on 'File' then 'Options'. Next, choose 'Add-Ins' from the menu at the left. Find and click once on 'Analysis ToolPak' in the list of Add-Ins. Press the 'Go...' button. Select the 'Analysis ToolPak' checkbox. Click on the 'Data' ribbon, a new option has appeared on the righthand side called 'Data Analysis'. Press this, and then find 'Regression' from the list. The 'Regression' window and the output table is the same as for Excel 2003.

For more complex graphs the uncertainty in the fitting parameters can be found by fitting the data to an equation using a least squares fitting method. This is illustrated in Section 4.18.5.

4.20 Fourier Transform

A Fourier transform allows a periodic signal consisting of more than one frequency to have its component frequencies and relative amplitudes identified. This can be useful in many areas of Physics where a periodic signal is measured. It is often implemented using a computer algorithm called a Fast Fourier Transform (FFT).

In Excel 2003, first check that the 'Data Analysis' command is on the tools menu. If it is not present, the Analysis ToolPak must be installed: see Section 4.19.

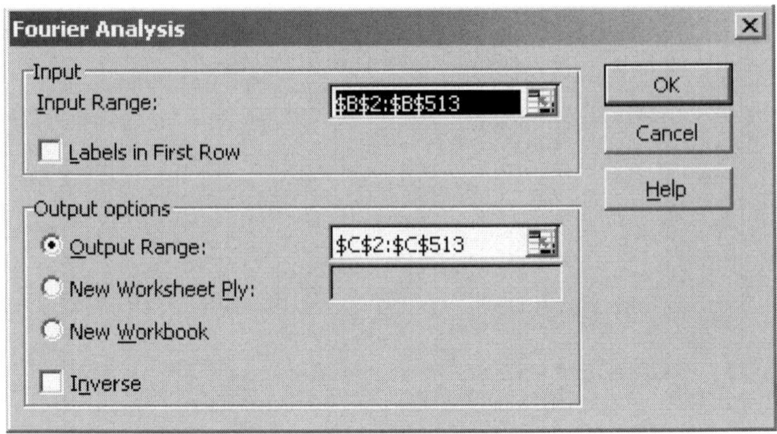

Figure 78: Entering the data ranges into Fourier Analysis window.

As an example, label five columns in Excel: 'Time', 'Data', 'FFT', 'FFT Frequency' and 'FFT Magnitude'. Next fill the time and data columns with the results from the experiment. In this example the time resolution is 0.01 and the data is the equation $y = \sin(10t) + 2\sin(5t)$.

A sometimes awkward limitation of the FFT algorithm is that the the number of pairs of data points must be equal to $N = 2^k$ where k is a whole number: for example $k = 9$, 10 or 11 gives $N = 512$, 1024 or 2048 points. The greater k, generally the greater the accuracy of the results of the FFT. To fit data to this constraint either remove some, re-sample it an appropriate number of times or pad it out with zeros.

To find the sample frequency take the total number of data points N and divide this by the difference between the times for the first and the last points $\Delta t = t_N - t_1$:

$$f = \frac{N}{\Delta t} \tag{129}$$

In this example $N = 512$, $\Delta t = 5.12$ and $f = 100$.

Next, fill the 'FFT Complex' column. In Excel 2003 choose 'Tools, Data Analysis, Fourier Analysis'. In Excel 2010 choose the 'Data' ribbon, then the 'Data Analysis' button and the 'Fourier Analysis' option. The 'Input Range' is column B, cells 2 to 513 so enter $B2:$B513. The 'Output Range' is $C2:$C513. See Figure 78. Column C now contains the complex number results of the FFT. The 'FFT Magnitude' in column E can now be calculated by multiplying the absolute value of the complex number (in column C) by 2 and

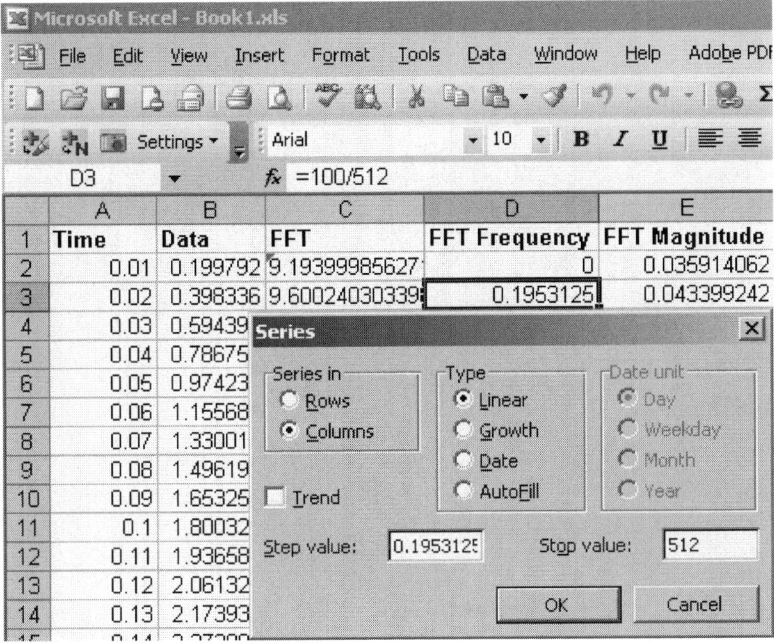

Figure 79: Filling in the FFT Frequency column

dividing by N. In this case $N = 512$ so the formula $= 2 \times 512 \times IMABS(C2)$ can entered into cell E2. This can be copied down to the remaining cells below by clicking a dragging the bottom right hand corner of the cell or using the key combinations Ctrl-Shift-End and then Ctrl-D.

Finally complete the 'FFT frequency' in column D. The general formula for this is p multiplied by the sampling frequency f divided by number of data points N. Where p is an integer starting from zero. That is: cell D2 is zero, D3 is f/M, D4 is $2f/N$ and D5 is $3f/N$ etc. Instead of manually filling in each of the rows, automatically fill the column using the Excel 'Series' function. Set D2 as zero and set D3 in this example as $= 100/512$. In Excel 2003 Click in cell C3 then select 'Edit, Fill, Series' to automatically fill the rest of the column. In Excel 2010 select the whole C column then from the 'Home' ribbon select the 'Fill, Series' option. In both versions now select 'Series in Rows' and then set the 'Step Value' as the result from cell D3 which is f/N and then set the 'Stop Value' as N: see Figure 79.

Finally plot a x-y scatter graph (with a line) of 'FFT Frequency' (x axis) and the 'FFT Magnitude' (y axis) in the usual way: see Figure 80. Do not plot

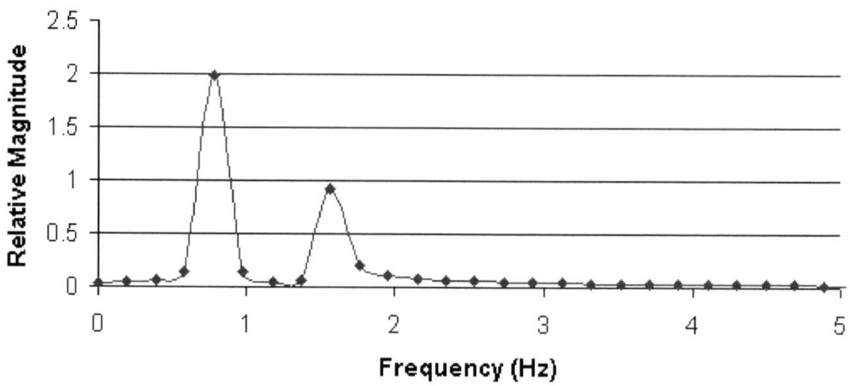

Figure 80: Graph of FFT Frequency against FFT Magnitude.

more than the first half (that is the first $N/2$) the rows as the FFT spectrum begins to duplicate itself.

In this example it can be seen that there are two frequencies present. The lower frequency at around 0.8 Hz has twice the magnitude of the higher frequency at around 1.6 Hz. This corresponds with the simulated data fed into the FFT from the equation $y = \sin(10t) + 2\sin(5t)$. The standard format of sine wave is $\sin(\omega t)$ where $\omega = 2\pi f$. Thus the $\sin(10t)$ term has a frequency of $10/2\pi = 1.59$ Hz and the $2\sin(5t)$ term has twice the amplitude and a frequency of $5/2\pi = 0.80$ Hz.

4.20.1 Window Functions

To improve the accuracy of FFT taken of experimental data, N could be increased and data covering more periods or a larger range of times could be used. If the data input to the FFT is not periodic (that is it does not contain exactly an integer number of periods of the wave) then leakage will occur in the Fourier transform spectrum potentially obscuring some frequencies or adding spurious ones: see Figure 81. This problem can be corrected by applying the appropriate windowing function to the data. The original data is multiplied by a window function which drops to zero at the limits of the data window. A number of windowing functions have been suggested each with advantages and disadvantages in relation to their accuracy at preserving the correct frequency and amplitude and to which signal type (sinusoidal or random) they are best

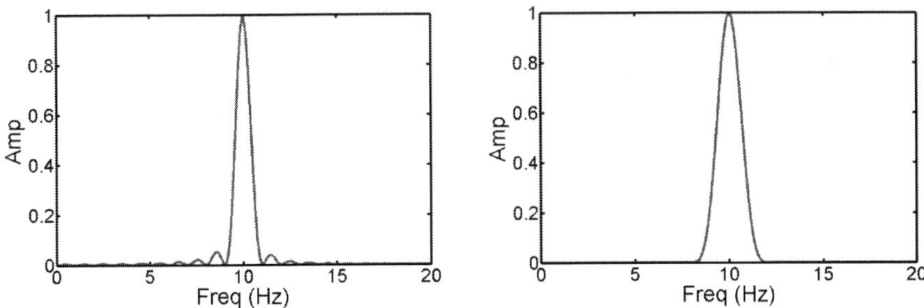

Figure 81: Fast Fourier transform of a sine wave with frequency 10 Hz. Left panel shows leakage with no windowing function present. Right panel shows leakage reduced with a windowing function.

applied. Examples include the Blackman (best at reducing spectral leakage), the Flat Top (best for amplitude accuracy) and None (best for frequency accuracy). A good all round function is the Hanning:

$$H(i) = 0.5\left(1 - \cos\left(\frac{2\pi i}{N-1}\right)\right) \tag{130}$$

This, or any other window function, can be applied by multiplying the i-th y data point by $H(i)$.

5 Presenting Results

This chapter looks at how results taken in the laboratory are recorded and presented to others. Each week, or for each experiment a script will be given which outlines what must be done and what must be measured. As students progress through a degree and gain more experience of work in the laboratory the scripts will become less detailed, leaving more room for them to work out the procedure themselves.

Almost certainly, results will be recorded and general notes taken about the experiment in a laboratory notebook. Students may be asked to produce a brief write up and complete the analysis for each experiment and hand this in. A small number of experiments are usually chosen and written up in more detail and students may produce a poster or even give a presentation on a particular experiment.

This section focuses in some detail on the layout and content of a laboratory notebook and formal report. It gives an overview of how a computer might be used to enhance a report and include special characters, equations, tables, graphs and references.

This section should be extremely valuable in giving an overview of recording and presenting results and hopefully pointing students in the correct direction for further investigation on the internet. A shortage of space means that it can not explain every detail of every software programme which may encountered, nor can it explain the specific details of how a particular university requires work to be produced.

5.1 Lab Book

While working it lab it is essential to keep a good record of everything done so that it may be possible to repeat the work from only these notes after a significant period of time has passed, perhaps a year or two. The following advice is generally considered good practice for keeping a lab book:

- Each experiment should begin on a new page with an underlined title and date.
- A brief description of the method being carried out is useful along with annotated diagrams or printed photos of the apparatus which show the procedure and exactly which measurements are being taken.

- All measurements should be recorded directly into the lab book. Initially, no mental arithmetic or calculator work should be carried out on them.
- Try to record all data in tables. Make sure a careful note is made of the name of the quantity being measured, its units and if possible an estimate of the error or uncertainty in the readings.
- Try to use appropriate units and powers of 10. Do not try to measure the work function of a metal in Joules, rather use 10^{-19} Joules or even eV in the table heading.
- Do not write on rough paper or copy pieces of data from one page to a neat copy later. Try to stick to one, original, reasonably neat and presentable table.
- Lab books should have crossings out and corrections throughout! If not it appears although they have been written up at a later date. This not only wastes time but can cause copying errors as well as potential loss of data written on loose sheets of paper.
- Take care to record the full filenames and locations of any data files stored on computer which need to be analysed at a later date. Often these will take the form of a name followed by sequential numbers. For instance mydata.000, mydata.001, mydata.002 or mydata000.txt, mydata001.txt, mydata002.txt. Remember to make a copy on a memory stick or email a copy before leaving the lab.

5.2 Reports

As well as keeping a neat lab book showing all the results and analysis of each experiment, students will often be required to write up two or three experiments formally as though they were scientific articles. This is a vital part of becoming a good Physicist, as once a cutting edge experiment is completed, the results analysed and a new discovery has potentially been made the results need to be communicated with other Physicists. This has traditionally been achieved by writing the experiment up as a scientific paper and then submitting it to a scientific journal. The journal then ask a number of other Physicists (the identities of whom are unknown the the authors) working in similar fields at other universities or research laboratories to be a referee for the paper and to make critical comments upon the work. Referee comments are passed back to the authors for their comment or so that they make modifications to their paper. Based upon these comments the journal editors will make a decision to accept or reject the paper for publication.

The aim of writing formal reports is to develop skills in writing papers: these will be vital in a future career in academia and almost certainly useful for a future career in any other field. They provide a good opportunity to gain a deeper understanding of an experiment beyond that achieved in a one or two day laboratory session.

This section gives an overview of the typical structure of a formal written report and provides some advice on how to include references and write a report on a computer.

5.2.1 Software

Simple reports with a small number of equations, figures and references can be most simply written in Microsoft Word, OpenOffice Writer or an equivalent computer programme. These programmes are convenient since they immediately reflect changes made to formatting and content of the report, making it easy to see how the report will appear when printed. However, numbering and referring to equations, graphs and tables is more tedious especially with a longer report. It can also at times be difficult to maintain consistent formatting and spacing throughout a long document and to insert special characters and complex equations. Microsoft Word can be obtained as part of the Microsoft Office suite of programmes from either an internet or high street shop. OpenOffice Writer is free and can be obtained from the OpenOffice website [17].

A further alternative to traditional software programmes is to use Google Docs which is a web based document creator and editor. Its advantage is that the document is stored on Google's servers, so can easily be edited at a number of computers and also jointly worked on by a number of people (useful for collaborative work). At the time of writing, Google Docs is sufficient for more basic work, although some of the more advanced features of Microsoft Office and OpenOffice are not yet present: for example different error bars for each data point can not be included on graphs.

More complex reports, perhaps those for final year projects or PhD theses are best written in LaTeX (pronounced Latek): in fact this book was written in LaTeX. LaTeX has a reasonably significant entry barrier and learning curve. It is written as a simple text document with all formatting, equations and figures inserted by using text based commands embedded within the text of the document. This makes it somewhat difficult to visualise a document without compiling it and then opening the resulting pdf file. The text based commands also need to be learnt or a guide referred to. However equations, special characters and references can easily be inserted and tracked and the layout precisely controlled. To start running LaTeX first install a TeX distribution such as MiKTeX [18]. It makes the compiling simpler if a text editor designed for use with LaTeX such as WinEdt [19] is used.

5.2.2 Structure

Write ups of experiments should provide enough information for a reader to understand the aims and outcomes of the experiments, the approach taken, the measurements made, the analysis conducted and the results obtained. They should be written in a formal style and the language used should be precise: the level should be such that it should be understandable by a peer who may not have any detailed knowledge of the specific experiment. The data and analysis should be presented in a manner that allows the reader to evaluate the significance of the results. The use of the words 'I', 'you' and 'we' should be avoided: the report should be written in an impersonal style i.e. the third person.

A formal laboratory report should be self-contained and structured so that it is easy to follow. A common approach is to break the report into a number of sections with clear headings. Sections should be numbered, as should diagrams, graphs and equations. This allows easy and precise referral during the text of the report: for instance to state that "using equation 2 the data shown in figure 3 can be used to plot the graph in figure 4". Bold fonts, underlining, numbered lists and bullet points can be used to make the report easier to follow for the reader.

A common structure which could be followed in most cases:

Title A short title which describes the experiment.

Abstract A few sentences (perhaps less than 100 words) which summarises the experiment and the results, giving quantitative information about the findings. The abstract should not contain references to the rest of the paper or other work.

Introduction The introduction sets the scene for the experiment. It should describe the background to the problem under investigation explaining why it is interesting and encompassing a critical view of other work on the same problem: what has been done and what still remains to investigate.

Theory A theory section should explain the physics behind the problem and show how the experiment carried out and the quantities measured can be related to the problem.

Experimental Methods A description of the apparatus used, including a diagram. This should include information on how the apparatus was built or connected together and any special features of the equipment or particular problems which had to be overcome. A description of the measurements taken and any special techniques which were required.

Results and Discussion Raw data, especially large tables of numerical data should be placed in an appendix at the end of the report. This section should show how the data was analysed to give a final result and contain any important intermediate results. A small number of graphs or summary tables are appropriate, although most extensive data sets can be put into an appendix. It is vital to include an estimate of uncertainties and to consider all sources of experimental error in the final results.

Conclusion The results of the experiments, together with appropriate uncertainties, should be summarised and compared with other values. The results should be set in context and their significance noted. Suggestions for possible future work can be made where appropriate.

References Sources of information such as other papers, articles and books should be listed here.

Appendix Detailed tables, graphs, derivations or computer programmes should be presented in the appendix.

In Microsoft Word or OpenOffice Writer the headings may be formatted to have a specific style: perhaps a larger font and in bold. In LaTeX simply use the 'section' commands as follows:

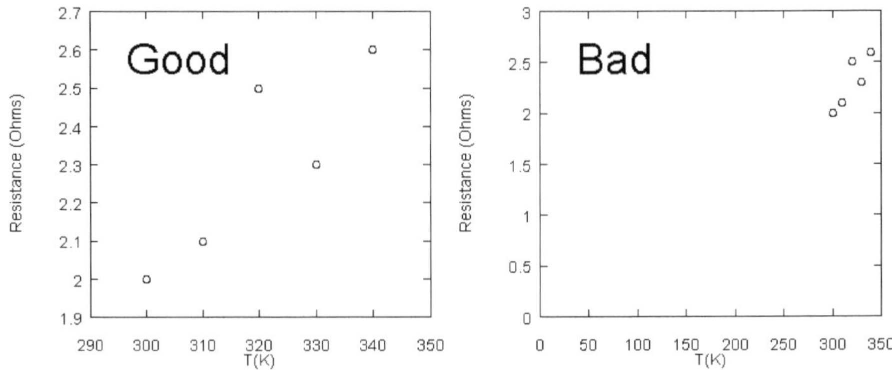

Figure 82: Good and bad scaling of axes.

```
\section{Section Title}
Some text.
\subsection{Subsection Title}
Some more text.
```

5.2.3 Graphs, Tables and Diagrams

Data should always be presented in graphs as they are efficient at showing any trends as well as scatter in the data. Graphs should always have titles, axis labels and units: make sure they are large enough to be read when the graph is scaled to fit into a report. Numerical values on axes should always increase in equal steps occupying equal distances on the page/screen (except for a log graph). However the values do not need to start from zero: for example an axis may be labeled 50, 60, 70, 80 but not 50, 55, 60, 70, 80. Axes should be scaled so that the data points on a graph occupy as large a portion of the graph as reasonably possible. For example see Figure 82 which shows good and bad scaling for the axes of a graph. Graphs should always be scatter graphs. Never draw line graphs where the points are joined like a dot-to-dot. If a computer generated fit line is used the parameters and equation should be clearly displayed. For example see Figure 83 which shows good and bad practice for adding lines to graphs. Finally, show x and y error bars where possible as illustrated in Figure 84.

An important skill of report writing is appropriate selection of data. Try to find ways to plot several sets of data in one figure. For example if there

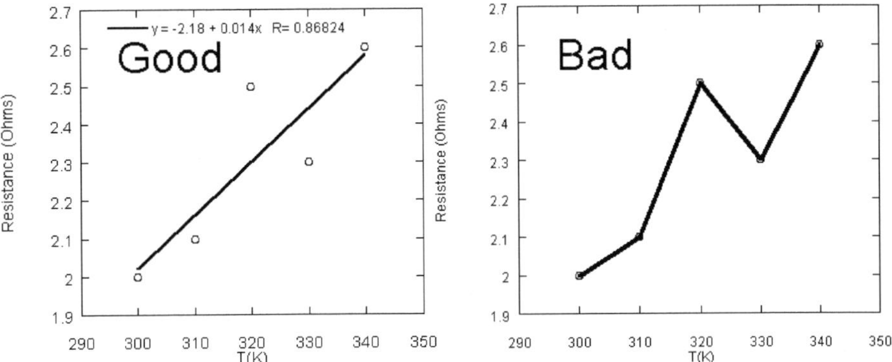

Figure 83: Good and bad lines on graphs.

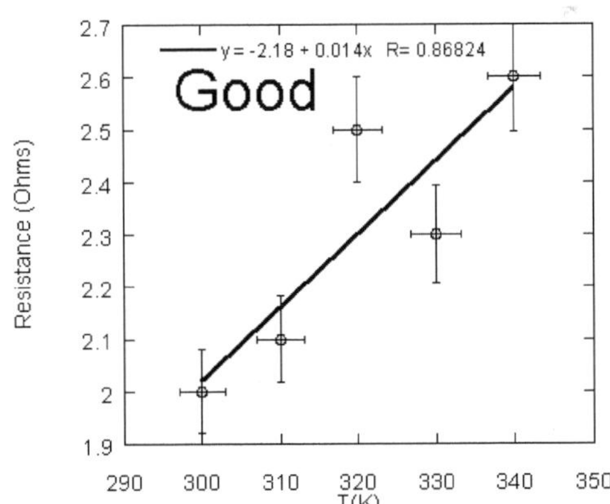

Figure 84: Always add x and y error bars where possible.

are several sets of data of resistance against length of a wire (perhaps with different cross sectional areas) these could be plotted on the same axes using different shaped markers (crosses, plus signs, triangles and circles) or different colours (only if the report is printed in colour). This will help distinguish the data sets. A legend should also be included which links the different markers or colours to the relevant cross sectional area.

Tables should be displayed with clearly labeled column headings, titles and units. In large tables it is usually essential to have grid lines, although these may be omitted in smaller tables.

Smaller graphs, diagrams or tables may be inserted into a larger figure: perhaps in a region of the graph where there is no data. All figures should have a self contained caption which explains what the figure shows. For example: "Figure 1: (electric) potential difference plotted against current. The filled squares are data using a carbon-zinc cell; the open squares are data from a lithium polymer cell. The curve is a fit to equation 3.".

All tables, diagrams and graphs should be discussed in the text and be relevant to the experiment performed. They should be placed within the text, close to the place where they are being discussed rather than collected at the end of the document.

Microsoft Word and OpenOffice Writer

Images which contain diagrams can be inserted in Microsoft Word 2003 by selecting the 'Insert, Picture, From File' menu and then choosing the correct image file. Tables can be copied from Microsoft Excel or inserted using the 'Table, Insert, Table' menu. Graphs are most easily copied from Microsoft Excel. In OpenOffice Writer the menu choices are identical.

LaTeX

In LaTeX images can be inserted using the following commands:

```
\begin{center}
\begin{figure}
\includegraphics [width=4in] {folder/filname.png}
\caption {Graph showing the change in A as a function of B.}
\label{graphname}
\end{figure}
\end{center}
```

The width of the image can easily be modified by specifying a number of inches or cm. The label command gives the graph a name so that it can be referred to automatically by its figure number elsewhere in the report by using the command:

```
\ref{graphname}
```

Graphs must be inserted as images. They may be screen captured from Excel using the Print Screen key on the keyboard. They can then be pasted into an image editor (such as Irfanview [20]), appropriately cropped and saved as an PNG image file. If using Kaleidagraph then a graph can be exported using the 'File, Export, PNG' menu option.

Tables may be inserted using the following commands:

```
\begin{table}
\begin{tabular}{ll}
\hline x & y=x$^2$ \\
\hline 1 & 1 \\
 2 & 4 \\
 3 & 9 \\
 4 & 16 \\
\hline
\end{tabular}\label{tablename}
\end{table}
```

These commands to insert images and tables mean that LaTeX will determine their position in the document to best suit the flow of test and figures across pages. If insertion at a fixed point in the text of preferred, regardless of how the text flows around, omit the begin/end figure or begin/end table commands.

To create bullet point lists in Latex, use the 'itemize' structure:

```
\begin{itemize}
  \item The first item
  \item The second item
  \item The third item
\end{itemize}
```

and the create numbered lists use the 'enumerate' structure:

```
\begin{enumerate}
  \item The first item
  \item The second item
  \item The third item
\end{enumerate}
```

5.2.4 Referencing

When writing a report always make clear which parts are written by the author and which parts are taken from, based upon, derived from or copied from other sources. For instance it is not possible to state "It is well known that the charge on an electron is 1.6×10^{-19} C". Statements like this should refer to a source which backs it up, for example: "The data in the present report give a value of the charge on an electron $e = 2.4 \pm 0.2 \times 10^{-19}$ C, which compares well with the study of Price and Arthur [18] who obtained $e = 2.2 \pm 0.1 \times 10^{-19}$ C".

Always check references carefully, especially the volume and page numbers and the spelling of authors' names/any unfamiliar words in the title.

There are two accepted styles for labeling references: the numerical system and the Harvard system.

Referencing: The Numerical System

The numerical system is generally in wider use in the sciences, especially Physics. It is the method adopted by most journals for the publication of original research. References are listed in a numbered list at the end of the paper. In the main text of the paper, the contribution of a reference in indicated with the relevant number either as a superscript[34] or in square brackets [34]. It is good practice, although not essential, for the references to be numbered in the order in which they are referred to in the text.

Various pieces of software can help with this and make the job of ordering numerical references less tedious. Microsoft Word has the option for endnotes. To insert a new reference choose 'Insert, Reference, Footnote', then choose the 'Endnotes' bullet point. This will give a list at the end of the document, rather than at the end of each page (if the 'Footnote' option is chosen).

In LaTeX a separate bibliography file can be created with the file extension '.bib'. In the '.tex' document the bibliography file is linked by including the following lines of code:

```
\bibliographystyle{unsrt}
\bibliography{file_name}
```

This 'unsrt' style produces a list in which the references are listed in the order in which they are referenced in the text. Changing the style to 'plain' gives a list in which the references are alphabetical by first author surname. The '.bib' file is a text file with a list of references. For example, the entry for one of the references used in this publication is arranged as follows:

```
@Article{liquidoxygen,
  title = {Making Liquid Oxygen},
  author = {French, M. M. J. and Hibbert, Michael},
  journal = {Phys. Ed.},
  volume = {45},
  pages = {221},
  year = {2010},
}
```

and the reference for a website is given by

```
@MISC{winedt,
    note = {http://www.winedt.com}
}
```

The list of references is inserted into the LaTeX document with

```
\bibliographystyle{unsrt}
\bibliography{references}
```

Referencing: The Harvard System

Most books and journals which predominantly publish review articles use the Harvard System of referencing. References are listed alphabetically by first author surname and are labeled by the Author surname and year of publication. In the main text of the paper, the contribution of a reference is indicated with the relevant surname and year e.g. [French 2012].

It is reasonably easy to keep track of Harvard style references in Word with a simple list of references at the end. In LaTeX a '.bib' file can be used as described in the numerical references section but the style must be set to 'alpha'. This gives a reference which includes the authors' initials and the publication year. More flexibility with the layout of these references in Word can be obtained by installing a special plugin and in LaTeX by using an additional specialised package.

5.2.5 Equations

Microsoft Word

In Microsoft Word 97 – 2003 equations can be inserted using the 'Insert, Object, Microsoft Equation 3'. This brings up a tool bar, from which Greek characters, mathematical expressions and layouts can be selected and inserted into an equation object. Figure 85 shows the tool bar being used to select a fraction arrangement.

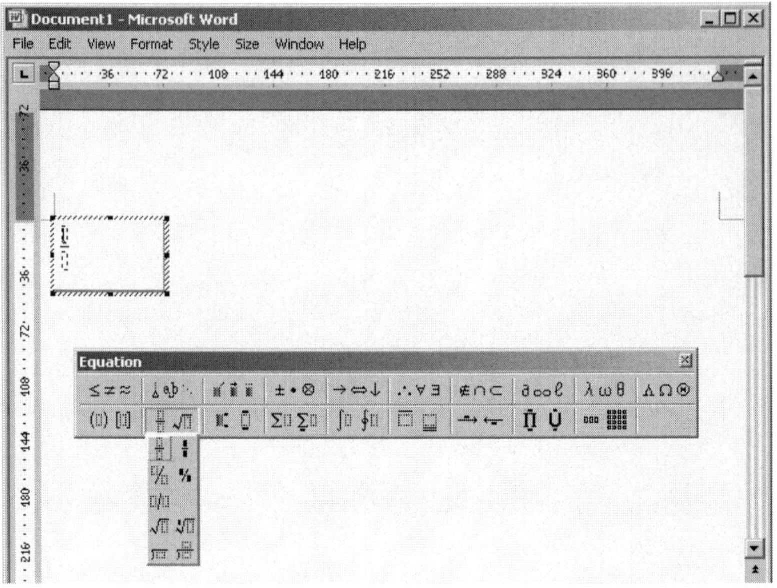

Figure 85: The Equation Tool Bar in Microsoft Word 2003.

In Microsoft Word 2010 equations can be inserted by choosing 'Equation' from the 'Insert' ribbon. See Figure 86. A further 'Equation Tools, Design' ribbon appears which gives a full range of options of mathematical expressions and layouts which can be inserted as shown in Figure 87. Greek characters are inserted by returning to the 'Insert' ribbon and choosing 'Symbol'. Then select 'More Symbols' and in the 'Symbol' window, the font should be changed to 'Symbol' as shown in Figure 88.

LaTeX

In LaTeX all equations are inserted by typing a text based command. For example fractions can be inserted by typing

```
$\frac{a}{b}$
```

which produces $\frac{a}{b}$ inline with the text or

```
\begin{equation}
\frac{a}{b}
\end{equation}
```

which produces

$$\frac{a}{b} \tag{131}$$

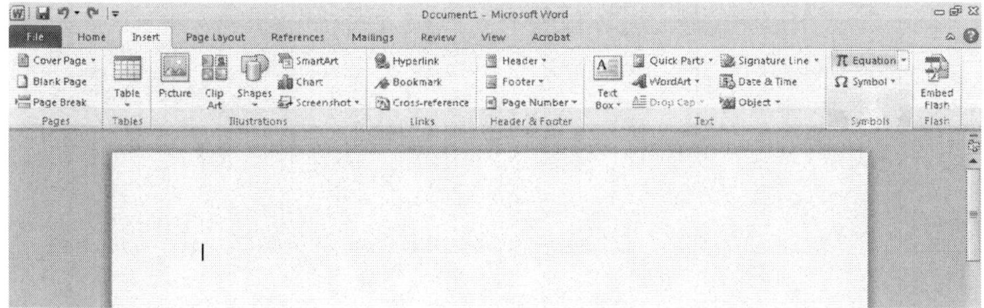

Figure 86: Inserting an Equation in Microsoft Word 2010.

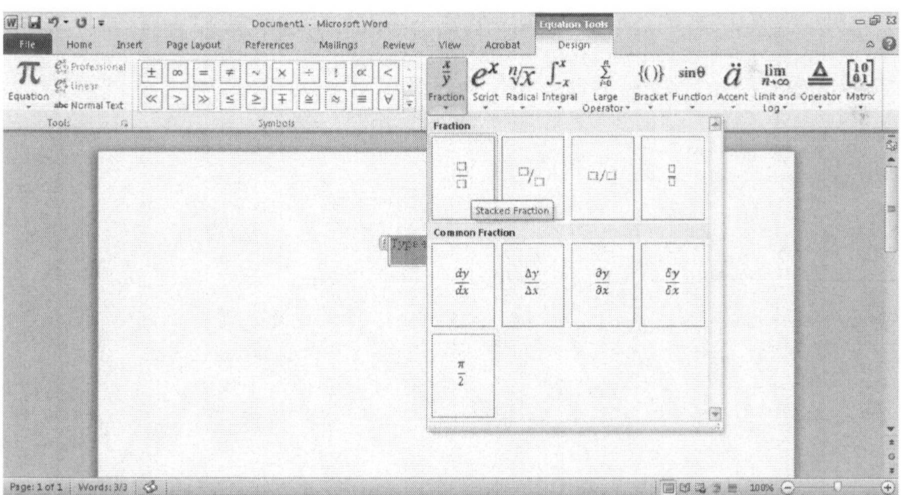

Figure 87: The Equation Tool Bar in Microsoft Word 2010.

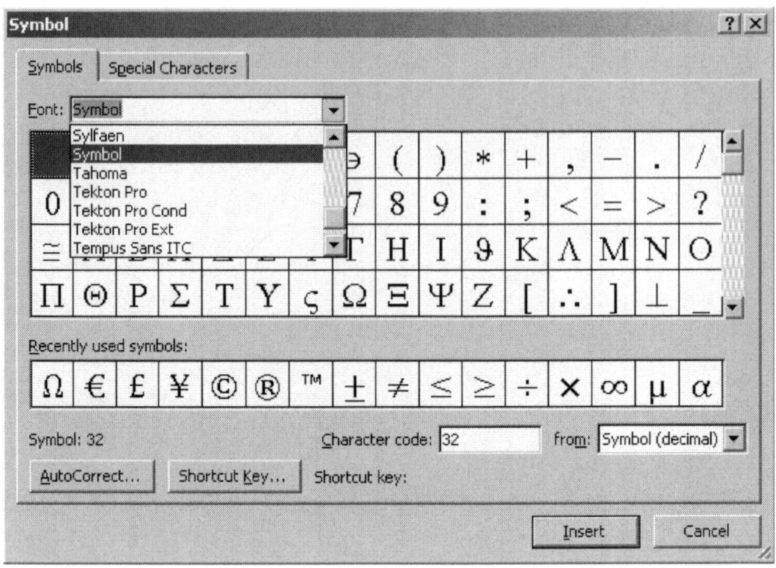

Figure 88: Inserting Greek characters in Microsoft Word 2010.

which is a numbered equation within the document. Brackets which automatically scale to the appropriate height are produced using the commands:

```
\begin{equation}
\left[ \left( x  \right)\right]
\end{equation}
```

which produces

$$\left[\left(x\right)\right] \tag{132}$$

To cover a number of other common functions consider the following example:

```
\begin{equation}
\int_0^\pi A\times\cos^2 \theta \partial\theta = A\times\int_0^\pi
\left(\frac{e^{\imath\theta}+e^{-\imath\theta}}{2}\right)^2
\partial\theta
\end{equation}
```

which produces

$$\int_0^\pi A\times\cos^2 \theta \partial\theta = A\times\int_0^\pi \left(\frac{e^{\imath\theta}+e^{-\imath\theta}}{2}\right)^2 \partial\theta \tag{133}$$

Other symbols can be found in Table 22 or looked up on the internet using a search engine. For example to find how to produce a much less than arrow

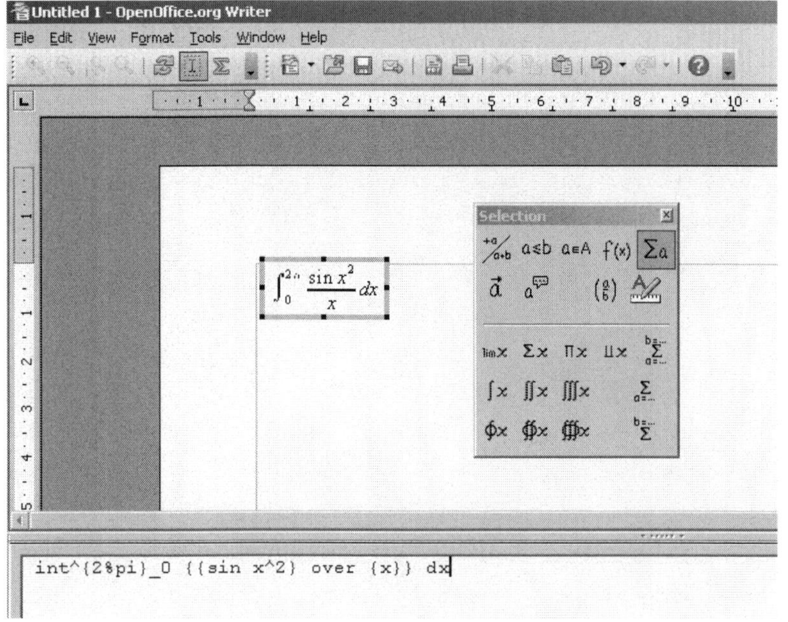

Figure 89: Inserting an Equation in OpenOffice Writer.

search for 'Latex much less than'. The insertion of Greek characters and subscripts in LaTeX are described in the next section on Special Characters.

OpenOffice Writer

In OpenOffice Writer an equation is inserted by selecting the 'Insert, Object, Formula' menu. This behaves as a hybrid between the text based method in LaTeX and point and click interface in Microsoft Word. Figure 89 shows the equation entry environment. Equations can either by typed into the text window at the bottom or entered by clicking in the 'Selection' box. Greek letters can be inserted by choosing 'Catalog' from the 'Tools' menu.

5.2.6 Special Characters

Often when writing reports or labeling tables and graphs characters will be needed which are not present on the keyboard such as a superscript 2, a degrees symbol or a Greek character.

In some software programmes it will be possible to turn on superscript or subscript options by finding a Format Text menu. However, this can be difficult or tedious in some applications. Much more simply a superscript 2 may be inserted by holding down the Alt key on the keyboard and then using the number pad on the righthand side of a standard keyboard (or by engaging the num lock function on a laptop) to type the combination 0178 and then releasing the Alt key. A superscript 3 may be entered by typing 0179 and the degrees symbol by entering 0176.

In some software programmes it may be possible to change the font to Symbol which allows a Greek character to be typed using the standard characters on the keyboard. A little trial and error will allow the selection of the correct character (although some common ones are a=α, b=β, m=μ, q=θ). Capital Greek letters can be produced by turning on Caps-Lock or holding the shift key, as usual.

Further number codes or characters maybe found and copied on an individual basis using the 'Character Map' programme. This can usually be found (well hidden!) under 'Programs' or 'All Programs' in the 'Start Menu' in the 'Accessories, System Tools' folder. Alternatively it can be run using the 'Run' command from the 'Start Menu' (or Windows button and R) and typing 'charmap'. Characters can be selected, copied and then pasted into an application.

In Microsoft Word, some special characters can be inserted by following the above instructions for inserting an equation.

In LaTeX, special characters are generally simple to insert. Greek letters are inserted by typing, for example:

```
$\theta$
```

which produces θ and a capital Greek letter is inserted by typing

```
$\Theta$
```

which produces Θ. Superscripts are inserted by typing, for example:

```
$x^3$
```

which produces x^3, but

```
$x^34$
```

produces x^34, where as

```
$x^{34}$
```

produces x^{34}. Subscripts are inserted by using the underscore character instead of the hat in a similar way.

There is a truly phenomenal array of symbols which can be inserted in to a LaTeX document. An internet search engine can be used to search for a general list of 'LaTeX special characters' or a more specific character by name. A brief list of some commonly used symbols is given in Table 22.

LaTeX code	Symbol	LaTeX code	Symbol
\alpha	α	\pm	$\pm$
\beta	β	\div	$\div$
\gamma	γ	\mp	$\mp$
\delta	δ	\otimes	$\otimes$
\epsilon	ϵ	\odot	$\odot$
\varepsilon	ε	\times	$\times$
\zeta	ζ	\le	$\le$
\eta	η	\ge	$\ge$
\theta	θ	\neq	$\neq$
\vartheta	ϑ	\ll	$\ll$
\iota	ι	\gg	$\gg$
\kappa	κ	\sim	$\sim$
\lambda	λ	\simeq	$\simeq$
\mu	μ	\approx	$\approx$
\nu	ν	\leftarrow	$\leftarrow$
\xi	ξ	\Leftarrow	$\Leftarrow$
\pi	π	\rightarrow	$\rightarrow$
\varpi	ϖ	\Rightarrow	$\Rightarrow$
\rho	ρ	\mapsto	$\mapsto$
\varrho	ϱ	\uparrow	$\uparrow$
\sigma	σ	\downarrow	$\downarrow$
\varsigma	ς	\ldots	$\ldots$
\tau	τ	\vdots	$\vdots$
\upsilon	υ	\cdots	$\cdots$
\phi	ϕ	\ddots	$\ddots$
\varphi	φ	\circ	$\circ$
\chi	χ	\infty	∞
\psi	ψ	\triangle	$\triangle$

\omega	ω	\hbar	$\hbar$
\Gamma	Γ	\ell	ℓ
\Delta	Δ	\prime	$\prime$
\Theta	Θ	\S	§
\Lambda	Λ	\partial	∂
\Xi	Ξ	\imath	$\imath$
\Pi	Π	\jmath	$\jmath$
\Sigma	Σ	\nabla	∇
\Upsilon	Υ	\%	%
\Phi	ϕ	\textbackslash	\
\Psi	Ψ	\	inter word space
\Omega	Ω	\,	short space

Table 22: Commonly used LaTeX symbols

5.3 Posters

It is common at scientific conferences for all attendees to be invited to produce a large A1 or A0 sized poster describing their recent work. These are displayed on boards, usually in a central location to give everyone a chance to view them and discuss the work. At university students might be given the chance to do this with an experiment they have carried out during their laboratory sessions, with a final year project or at the beginning of a PhD. It is likely that some guidance will be given on content, layout and length. However, a good outline would begin with a title, a list of authors and an abstract. The number of words should be kept to the minimum needed to give an introduction to the work, explain the experiment(s) carried out and give some analysis of the results and a brief conclusion. The emphasis should be on the inclusion of diagrams of the experimental setup where appropriate and graphs or other way of displaying the results. It is crucial to ensure that text is printed in a font size that is large enough to read: text should be not smaller than 28 point (perhaps 24 point for captions or references).

There are two examples of posters included in this book. Figure 90 was produced at the end of a final year research project and Figure 91 as a summary of an experiment undertaken during the early stages of a PhD.

5.3.1 Software

Posters can be produced in a programmes such as Microsoft Word, OpenOffice Writer or LaTeX, however it is more straightforward to use Microsoft Powerpoint or perhaps OpenOffice Impress. A single slide can be used, but the page size should be enlarged to the appropriate size. This is preferential to enlarging a normal slide as font sizes will retain their familiar sizes. Usually posters will be printed by an IT or printing department at the university: large poster printers are still expensive to buy and run and there may only be one per department. The paper will usually be a role of a certain width. Posters may be either portrait or landscape, but to avoid distortions when printing it is useful to find out the paper width or maximum printing width before starting a poster. Then either the height (landscape) or width (portrait) of the poster should be set to this.

5.3.2 Paper Sizes

Standard European paper sizes follow the A series with A0 being the largest, decreasing in size to beyond A6: see Table 23. There are three rules governing this series:

1. A0 paper has an area of 1m^2
2. The length divided by the width is $\sqrt{2} = 1.4142$
3. The shorter side of size A(n) is the long side of A(n+1) (that is A(n+1) is A(n) cut in half parallel to it shorter side).

Sizes of paper larger than A0 are not defined as clearly, but can follow the pattern in Table 23. There is a less common series called the B series. The size for the length or width is given by:

$$B(n) = \sqrt{A(n) \times A(n-1)} \tag{134}$$

Tolerances for the paper size are ± 1.5 mm for lengths up to 150mm, ± 2mm for lengths between 150 and 600 mm and $\pm$ 3mm for any dimension above 600 mm.

In Microsoft Powerpoint 2003 the page size is changed by selecting 'Page Size' from the 'File' menu and changing the 'Width' and 'Height' settings. In OpenOffice Impress the page size is changed by selecting 'Page' from the 'Format' menu and changing the 'Width' and 'Height' settings.

4A0	2378 x 1682 mm
2A0	1682 x 1189 mm
A0	1189 x 841 mm
A1	841 x 594 mm
A2	594 x 420 mm
A3	420 x 297 mm
A4	297 x 210 mm
A5	210 x 148 mm
A6	148 x 105 mm

Table 23: Papers Sizes.

Finally, it should be ensured that images are of sufficient resolution that they will not appear 'blocky' and distorted when they are printed out on a poster. Ideally there should be at least 150 dots per inch and probably less than 300 dots per inch to avoid excessively large file sizes. So if a graph is required to appear about 6 inches wide (15 cm) on the final poster, it should be between 900 and 1800 pixels wide.

5.4 Presentations

Research scientists also regularly give presentations/talks/seminars on their research work. Most universities make giving presentations a compulsory part of the undergraduate degree. Topics range from focusing on an experiment done in the laboratory, a final year project/disertation topic or on a certain topic set as a reading assignment.

Many people are nervous about or dislike giving presentations especially to large audiences. This is something which significantly improves with practice: try to take opportunities to practise speaking in front of people. If presentations are well prepared and have been practised to friends or in front of a mirror they become easier.

Common mistakes with presentations are writing a full script and trying to read it. This means that there is constant looking down at a script and not looking at the audience. Do not rush: speak slowly and clearly so that everyone can understand what is being said. Try to stand somewhere so that it is easy to engage with the audience. Too many people stand behind a table holding notes and appear to the audience to be 'hiding' or seeking protection

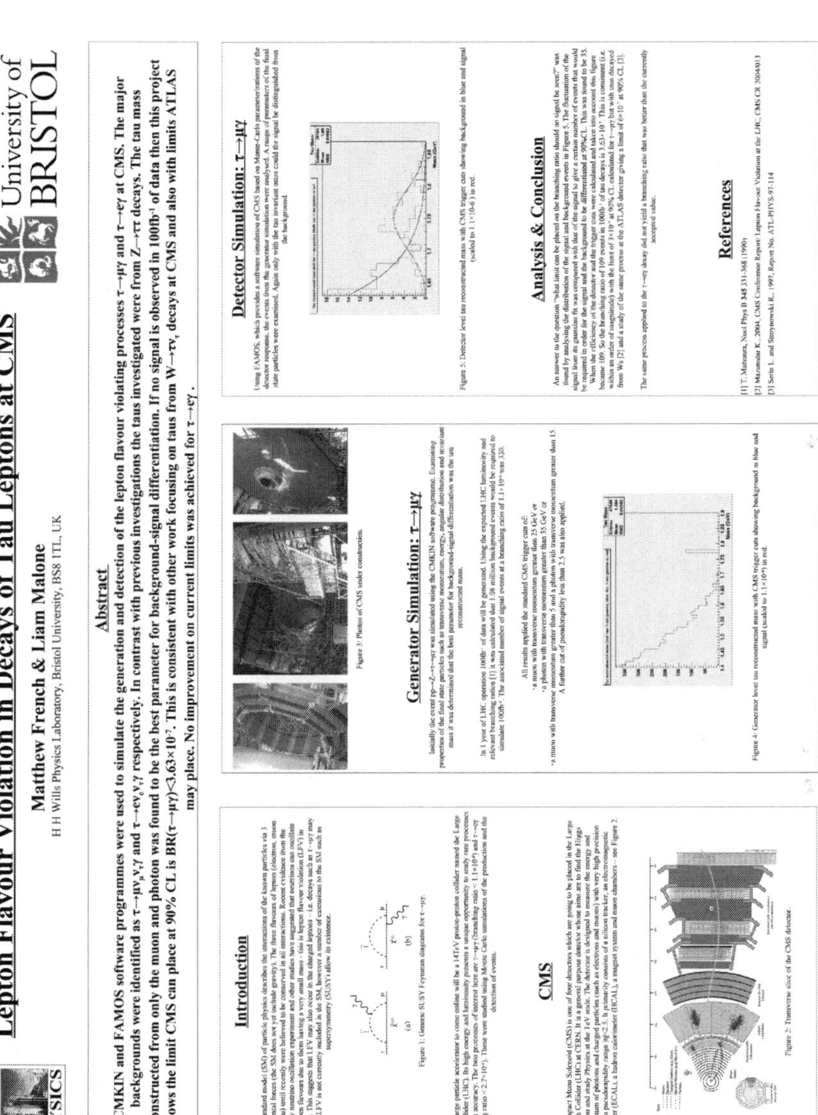

Figure 90: Example poster 1. Based on a final year project.

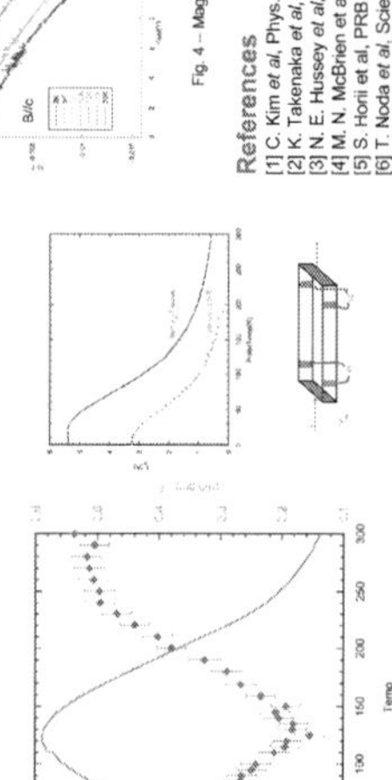

Figure 91: Example poster 2. Based on early work towards a PhD.

behind furniture.

Slides should not be blocks of text, perhaps a maximum of 30 words per slide as a general guide, nor should they have more than one graph per slide. Elaborate slide transitions and flashing text should always be avoided. Make sure text is large enough to read (this includes graph axis labels) and the colour of the text and background aid rather than hinder its legibility. For example, black on white is easy to read, black on dark blue or yellow on white is very hard to read. For ease of reading fonts should not be so complex and fussy that the audience struggle to read them, although recent research suggests [21] that if serif fonts such as Times New Roman or Monotype Corsiva are used, the audience will remember more of the presentation.

Start with a title slide. This should be followed with a slide giving a brief outline of the talk. A good way to plan a short presentation is to complete the outline slide and then turn each point from the outline into a slide title. Finally, include an acknowledgements and references slide unless this information is included elsewhere. As a rough guide around one slide per minute is needed, so a 10 minute presentation should have 10 slides in addition to slides for the title, outline, references and acknowledgements.

Figure 92 shows an example set of slides for a short presentation. The accompanying notes are as follows:

1. Originally a Sci-Fi creation, which was first proposed scientifically by Carl Sagan in 1961.
2. It hasn't been proven that humans can live in enclosed environments for long periods of time i.e. space stations.
3. Correct Temperature - not too hot (velocity of particles is greater than escape velocity of planet) and not too cold (can't heat it enough); Ability to retain an atmosphere - size of planet - enough gravity; Location of planet - solar winds remove atmosphere; magnetic field existence/strength; Within easy travelling distance of earth - transport of equipment, other resources if needed, people; Availability of resources - things needed to terraform - lots of water and CO_2; Mars - close, once had water flowing on surface, probably has frozen water at north pole and CO_2 at south pole, has days unlike the moon, days of the similar length, has seasons.
4. To start terraforming - first stage is raise the temperature of planet

by an estimated 4 – 20 K to melt southern cap of CO_2 and start 'run-away' greenhouse effect. Methods to do this include reducing albedo by covering polar caps with carbon or other black material to increase absorption of heat, using large orbiting mirrors to redirect sun light, the Russians have already built and tested in Earth orbit a 20 m space mirror in the Znamia Project, Using SF_6; CFCs no good since they are destroyed by UV light which breaks the C-Cl bond - Mars has no ozone layer to reduce UV light so CFCs don't stay around for very long - would require rapid replacement - PFC's have no C-Cl bond so would be much more long-lived - very efficient since only a few ppm would be needed to provide a significant warming, they are non-toxic, have long life times and could easily be made on the Martian surface. Ammonia containing asteroids - ammonia is a potent greenhouse gas, also the rock of the asteroid would contain a great deal of CO_2 Start of runaway greenhouse effect - CO_2 melts and is released to atmosphere from south pole, CO_2 also evaporates from surface rock, planet warms up, more CO_2 melts, planet warms up more This is a proven method to warm up a planet - it's what's happening here on Earth and should take approx 100 years to reach stability. A 4°C initial temp rise would eventually produce a 55°C temp increase. Human efforts would set the ball rolling, but nature would perform 99% of the work.

5. Atmospheric pressure increased - evaporation of CO_2 increases pressure; Water vapour in atmosphere - Mars needs a great deal of water transferred into vapour in the atmosphere - equivalent to a slab of ice 350 miles in diameter and 100 yards thick, this also helps to raise the temperature; Microbes and Plants introduced to increase O_2 level - microbes, especially genetically engineered or selectively breed have to be used first since they are very hardy, plants wouldn't survive in the initial atmosphere of almost all CO_2, algae and plants can be introduced later when there is slightly more O_2 approx partial pressure of 20 mbar, to help create oxygen from the CO_2. The content of N_2 would also have to be increased.

6. Significant increase in atmospheric pressure - currently Mars is 7 mbar, Earth is around 1013 mbar, with the evaporation of all the CO_2 in the polar caps and locked in rocks on the surface and before the introduction of microbes and plants the partial pressure of CO_2 could reach around

500 mbar; Liquid water available - melting of polar caps and water under surface due to temp and pressure rises UV and cosmic radiation decreased - excess oxygen can form into ozone (O_3) in the atmosphere, and protect the surface from UV radiation Danger that Mars becomes too hot and enter a quasi-stable high temperature regime.

7. References:
   http://www.users.globalnet.co.uk/ mfogg,
   Space 2000, Harry L Shipman, Plenum Press (1987)
   The Road to the Stars, Iain Nicolson, William Morrow and Co (1978)
   www.nasa.gov

5.4.1 Software

To create an electronic presentation use software such as Microsoft Powerpoint, OpenOffice Impress, Prezi or Google Docs. Powerpoint and OpenOffice produce conventional presentations with a series of slides which are stored on a computer. The Google Docs web based software has a 'Slides' application which can be accessed online from http://docs.google.com is very useful when working on a group presentation as it can be shared with other group members allowing them to access and modify the presentation. It also makes working on the presentation easier from more than one computer. More and more universities are now switching over to cloud based software like this. Finally, for a slightly different take on a presentation try Prezi[22]. This produces a presentation a little similar to a mindmap which is navigated around in a defined order.

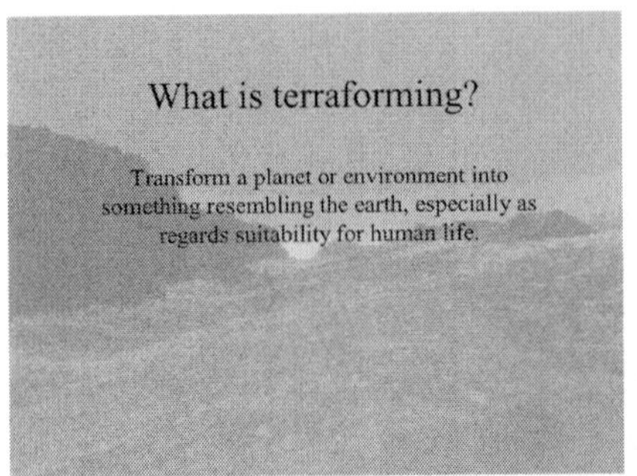

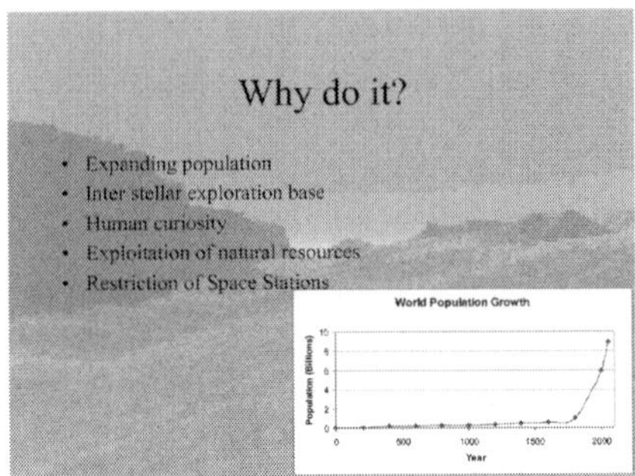

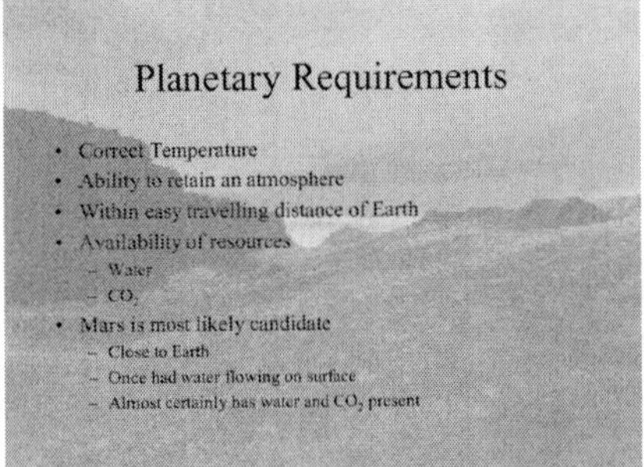

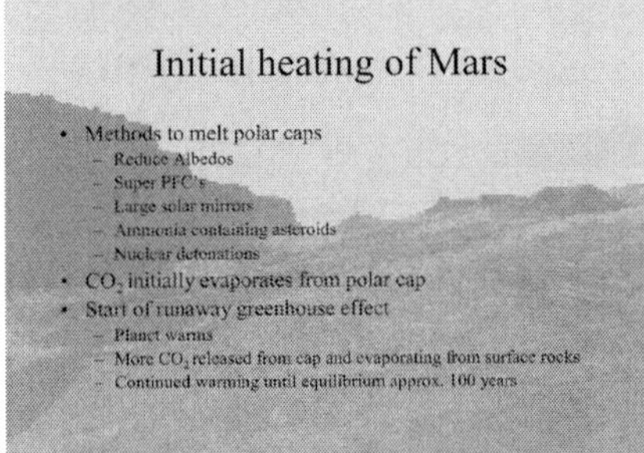

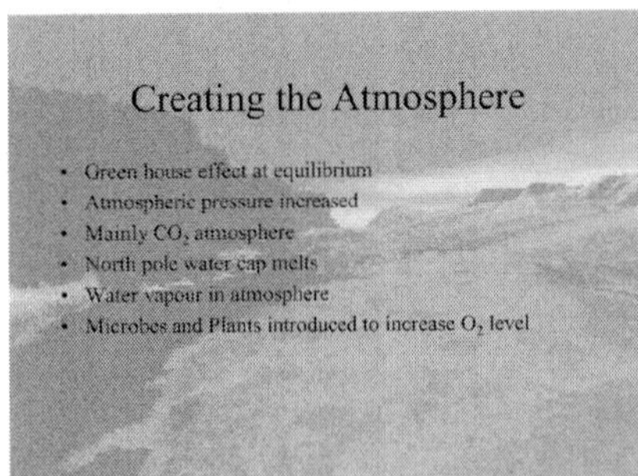

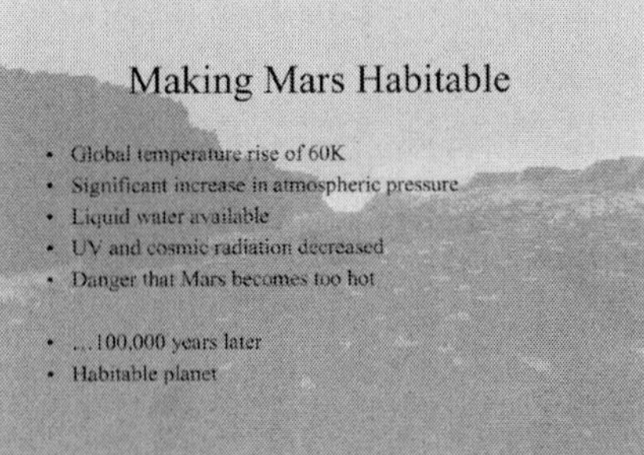

Figure 92: Example presentation slides

6 Common Experiments

The set of experiments completed in the first year of an undergraduate physics course aims to develop experimental skills and build confidence with new apparatus. Sometimes an experiment's main goal is to develop ability to carefully record data and apply the analysis techniques described earlier in this book. A number of universities start with the measurement of the acceleration due to gravity, g. Whilst the value of g should be familiar to every student, the goal of this experiment is to measure it to one part in 1000. This experiment trains students to calculate the uncertainty and error in their work and find ways of minimising them. It is common for experiments to then focus on the measurement of physical constants.

Over the course of the first year in the laboratory students will develop the ability to carefully read and follow written instructions, apply the data analysis techniques above to carefully recorded data, learn to use new apparatus, research and write formal lab reports and hopefully enjoy experimental physics.

6.1 Example Experiments

This section gives two example laboratory scripts. The first looks at the measurement of the acceleration due to gravity using a pendulum and the second looks at two methods to measure the speed of light. Some scripts, or parts of scripts will be described in detail giving you specific guidance about the measurements to make and the equations needed to complete the analysis. Others, which are perhaps designed for times when students have had prior experience of similar work, will only give an outline of what should be measured and suggest how a derivation be started to find the equations needed for the analysis.

6.1.1 Measuring Acceleration due to Gravity

This experiment is often the first which you will undertake in the undergraduate laboratory. The aim of this experiment is to measure the acceleration due to gravity to a precision of one part in a thousand. It also has the objective to teach you to understand errors. You will learn how to treat random error in measurements of the time period of the pendulum and how to combine errors from different sources to give an estimate of the error in the final value of the acceleration due to gravity.

You will use a rigid pendulum pivoted about a knife edge. For small oscillations the period, T, of the pendulum is given by:

$$T = 2\pi\sqrt{\frac{I}{mgh}} \tag{135}$$

I is the moment of inertia of the pendulum about the axis of rotation along the knife edge, m is the mass of the pendulum and h is the distance from the pivot to the center of mass of the pendulum. The moment of inertia is the rotational equivalent of mass. If we consider a simple pendulum (that is, a point mass on a a thin wire) which has $I = mh^2$ then Equation 135 reduces to the more familiar $T = 2\pi\sqrt{h/g}$.

Rearranging Equation 135 gives:

$$g = \frac{4\pi^2 I}{mhT^2} \tag{136}$$

In order to measure g to 1 part in 1000, you need to consider errors in I, m, h and T. Since T appears as a square term in Equation 136, you need to find T to 1 part in 2000. The errors in I, m, h are not independent since I is related to m and h, so the error may cancel to some extent. When considering random errors, it is important that you don't miss the possibility of systematic errors being introduced by your measurement process or technique.

Finding T to 1 part in 2000

The pivot is first clamped to the desk and the rigid pendulum is suspended from its knife edge. Care should be taken to ensure that the pendulum is swinging properly from the knife edge so that there is no sideways oscillation or movement.

Equation 135 is valid if the oscillations are small as it depends upon the approximation of $\sin\theta \approx \theta$ which only true for small θ. Therefore you should keep the amplitude of oscillation to a minimum: perhaps below ± 20 mm.

The pendulum should be set in oscillation and the time period, $5T$, for five complete oscillations should be measured using a stopwatch. A marker should be used in your view below/behind the pendulum to define a fixed point to start and stop the stopwatch. To minimise the error, this should be in the middle of the swing as it is here the pendulum is traveling the fastest. This means there will be the smallest ambiguity in the time at which the pendulum passes this point. Six readings x_i of the period $5T$ should be measured. Then calculate:

1. The sum of the six readings $\Sigma_{i=1}^{6} x_i$
2. The mean period, $\overline{5T}$, by dividing the sum of the periods from step 2, by six
3. The six differences $(x_i - \overline{5T})$ between the measurements and the mean $\overline{5T}$
4. The sum of the square of the differences $\Sigma_{i=1}^{6}(x_i - \overline{5T})^2$
5. The standard error in a single measurement $\sigma = \sqrt{1/5 \times \Sigma_{i=1}^{6}(x_i - \overline{5T})^2}$
6. The standard error on the mean of $5T$ $\sigma_m = \sigma/\sqrt{6}$
7. The standard error in the period, T, by diving the standard error in the mean of $5T$ by $\sqrt{5}$.

You will find that you are probably some way off achieving an error of 1 part in 2000 or 0.0005. You could, of course, continue to repeat the 5 period method enough times as the standard error in one period decreases as the number of repeats, n, increases. However due to the $1/\sqrt{n}$ dependence a very large number of repeats will be required.

An alternative approach is to reduce σ_m by reducing σ rather than increasing n. You can do this by increasing the number of periods measured from 5: this will reduce the random error, but also the systematic error in the period T.

If your reaction time is around ± 0.1s then to achieve 1 part in 2000, you will need to measure for around 200 seconds. To ensure you reach the target of 1 part in 2000, try to measure a time of around 300 – 400 seconds. However it will be very easy to lose count of the number of periods in such a long time so try the following method.

Using the stopwatch time the period of an integer number of oscillations (it doesn't matter exactly how many) which take around 60 seconds in total. The actual number of periods can be found by using your previous value of T and knowing there was an integer number of periods. You should now be able to fill in the table below:

Estimate of T, from above	
Time for N swings	
Implied value for N	
Improved estimate for T	
Current estimate of error in T	

Using your value of the error in T, you should find the error in your calculated value of N. This should be much less than 1, otherwise you will need to need repeat this step timing for a period shorter than 60 seconds.

You should repeat the procedure, but timing for around $300-400$ seconds and fill in the table again:

Estimate of T from timing for ≈ 60 seconds	
Time for N swings	
Implied value for N	
Improved estimate for T	
Current estimate of error in T	

You should now have a value for T with an error of less than 0.0005.

Finding h and m

These should be relatively straightforward. The mass m can be determined by placing the rigid pendulum onto a balance. The calibration of the balance should be checked using standard masses and a number of repeats taken. The mass and the error in the mass should be recorded. If the pendulum has mass around 1kg, then a measurement to the nearest 10th of a gram will give an error of one part in 10,000. To measure the height h the center of mass of the pendulum must be found by balancing the rigid pendulum on a knife edge. The pendulum rod can then be marked at the balance point. The measurement h should be taken from the knife edge at the top of the pendulum rod to the mark you have just made. Over a distance of around 500 mm, you will need to measure to the nearest 0.5 mm to achieve an error of less than 1 part in 1000.

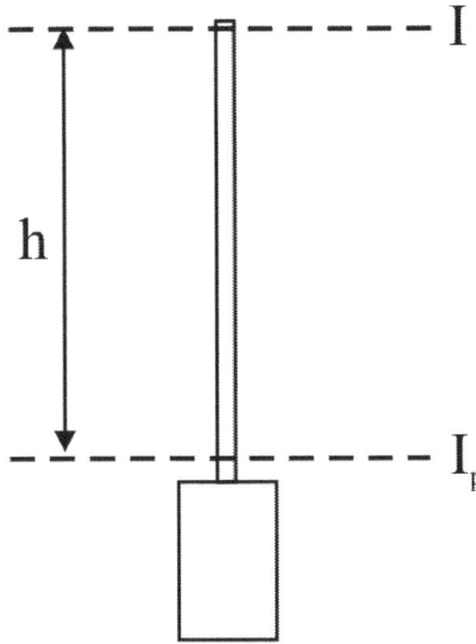

Figure 93: Moment of inertia of a pendulum.

Finding the Moment of Inertia

The final measurement to be taken is the moment of inertia, I, of the pendulum about the knife edge. With a compound object like the rigid pendulum, the moment of inertia can be most easily found by using the 'parallel axis' theorem. This states that the moment of inertia, I, of an object about a given axis is related to its moment of inertia, I_p, about a parallel axis through the center of mass by:

$$I = I_p + mh^2 \tag{137}$$

where m and h have the same definitions we have already used. See Figure 93. You should find that I_p is only about a tenth of the magnitude of mh^2 and so only needs to be measured to a precision of about 1 part in 100.

I_p, the moment of inertia of the pendulum about the axis through its center of mass, can be found by comparing torsional oscillations of the pendulum with those of a regularly shaped rectangular bar:

$$\frac{I_p}{I_b} = \left(\frac{T_p}{T_b}\right)^2 \tag{138}$$

where I_b is the moment of inertia of the bar about its center of mass, T_p is the time period of the torsional oscillation for the pendulum and T_b is the same for the bar.

First, find I_b. This is relatively simply as it is given by a standard formula for a regular rectangular bar:

$$I_b = \frac{m_b}{12}\left(a^2 + b^2\right) \tag{139}$$

where a and b are the length and width of the bar and m_b is the mass of the bar. Since you are looking for I_p to 1 part in 100, m_b is needed to 1 part in 100 and a and b are needed to 1 part in 200.

You should suspended the pendulum from a wire about its center of mass. The pendulum can then be displaced so that it undergoes a torsional oscillation. This means that it twists from side to side in the horizontal plane, with the center mass staying directly beneath the point of suspension. The time period of the torsional oscillation should be measured following a similar procedure to that used previously. This is repeated for the bar. Since you are aiming for an error in I_p of less than 1 part in 100, the time periods must have an error of less than 1 part in 200.

To determine the errors in I_b and then I_p do the following. Write $a^2 + b^2 = X$ then:

$$(\Delta X)^2 = 2(a\Delta a)^2 + 2(b\Delta b)^2 \tag{140}$$

and calculate ΔX. You may like to try proving this using the equations given in Section 4.15 of this book. The error in I_b can now be found from:

$$\left(\frac{\Delta I_b}{I_b}\right)^2 = \left(\frac{\Delta m_b}{m_b}\right)^2 + \left(\frac{\Delta X}{X}\right)^2 \tag{141}$$

The error in I_p is found using:

$$\left(\frac{\Delta I_p}{I_p}\right)^2 = \left(\frac{\Delta I_b}{I_b}\right)^2 + 4\left(\frac{\Delta T_p}{T_p}\right)^2 + 4\left(\frac{\Delta T_b}{T_b}\right)^2 \tag{142}$$

Finding g and the error in g

Since:

$$g = \frac{4\pi^2(I_p + mh^2)}{mhT^2} \tag{143}$$

The equation for the error in g will be somewhat cumbersome. If $\Delta I_p/I_p < 0.01$ then the error may be small enough to be ignored and a simpler equation may be derived to calculate the error in g. For the purpose of the error calculation:

$$g \approx \frac{4\pi^2 h}{T^2} \tag{144}$$

so:

$$\left(\frac{\Delta g}{g}\right)^2 = \left(\frac{\Delta h}{h}\right)^2 + 2\left(\frac{\Delta T}{T}\right)^2 \tag{145}$$

If $\Delta I_p/I_p > 0.01$ then a more complex equation must be derived for the error in g.

Discussion

Finally, it is worth identifying the largest sources of error in the experiment and writing a discussion focusing on whether the error in g met the target of being less than 1 part in 1000. You should consider how the value of g could be improved either by modification of the experimental technique or by redesigning the method completely. Systematic errors also need to be considered: is your location on earth or height above sea level likely to influence the result significantly: by how much?

6.1.2 Measuring the Speed of Light

The aim of this experiment is to make measurements of the speed of light by two different techniques. Firstly using a modern method, sending pulses of light down optical fibres and measuring the delay time with an oscilloscope and secondly using a method developed by Foucault in the 19th century. It will give you practice at using a range of apparatus, implementing precision measurement techniques and considering the safety risks associated with the use of lasers.

Optical Fibre Method

In this method you will measure the time taken for photons of visible light to propagate down various lengths of optical fibre. However, you can't make a simple speed equals distance over time calculation for one fibre as there is a time delay t_0 due to the electronics generating and detecting the photons which is not known. The total time taken for a photon to propagate from the transmitter to the detector can be given as:

$$t = t_0 + \frac{nx}{c} \tag{146}$$

where n is the refractive index of the optical fibre, x is the length of the optical fibre and c is the speed of light. If a series of measurements of t for fibres of different lengths are performed, the gradient of a graph of t verses x can be used to find a value for c.

To begin, connect one of the shorter cables between the light emitting diode (LED) transmitter and the photodiode fibre optic receiver. Make sure the LED transmitter is connected to the power supply and turn it on. So that pulses of light are produced, connect the TTL output from a signal generator to the input on the LED transmitter.

Next connect the signal generator to the CH1 input of a high frequency oscilloscope. Note that a normal oscilloscope won't be able to sample often enough to resolve the very short time differences you will find. Connect the fibre optic receiver to CH2 input of the same oscilloscope.

Setup the oscilloscope so that you can see two traces showing square waves on the display. The traces should be enlarged so that you can see one or two periods of the square wave. The CH2 trace should rise at a slightly later time than the CH1 trace, corresponding to the time delay due to the electronics and the length of the optical fibre between the LED transmitter and the fibre optic receiver.

You may need to adjust the position of the ends of the fibre optic cables to optimise the amplitude of the received signal, especially with the longer cables where the detected signal is weaker. The cursor controls on the oscilloscope can be used to measure the time delay. Choose the most appropriate points to measure between and try to justify your choice.

You should consider how reproducible your result is and whether or not it should be repeated. Finally, repeat the measurement with a range of optical fibres up to 100 m in length.

To analyse these results, plot a graph (including error bars for both quantities) of the time delay verses the length of cable. Use the graph to determine the delay time due to the electronics and the ratio n/c. The refractive index, n, will be given to you, allowing you to calculate c. Using multiple lines on the graph, estimate the error in n/c and use this to express the error in c. How many standard deviations is your result away from the defined value? Consider the effect of systematic errors: where might they arise, what could be done to minimise them and would they lead to an increased or decreased value for c?

Foucault's Method

A beam of laser light is focused using Lens 1 and then passed through a beam splitter. Part of the light enters the microscope and part makes its way to the rotating mirror. When light is incident on the rotating mirror it is reflected towards a fixed, spherical mirror. The fixed mirror always reflects the light back along the same path, returning it to

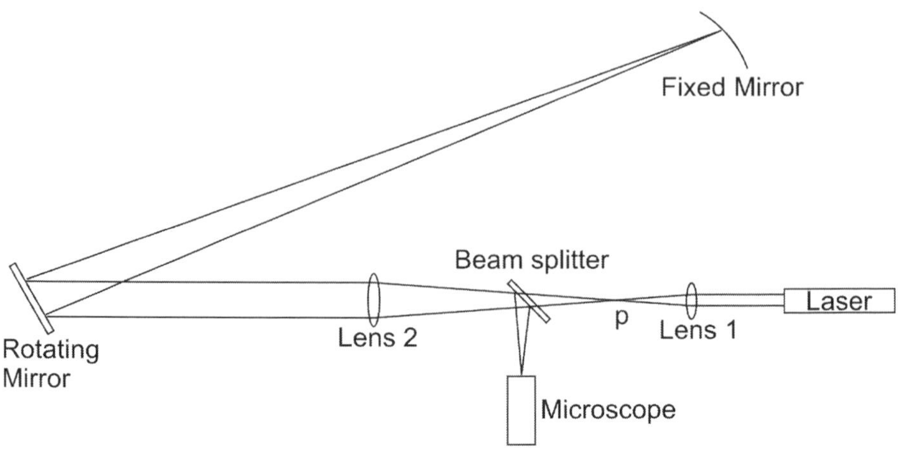

Figure 94: Faucault method diagram.

the rotating mirror and then the beam splitter. Some will be directed towards the microscope where it can be viewed in conjunction with the light which traveled the shorter route. A diagram of the apparatus is shown in Figure 94.

First, lets consider the point of reflection on the fixed mirror as the angle of the rotating mirror changes. Figure 95 shows the path of a light ray which is reflected from the rotating mirror at angle 2θ. At a slightly later time, the rotating mirror has moved and now the light ray reflects at angle $2(\theta + \delta\theta)$. This means the two light rays reflect from the fixed mirror at different points. If the distance between the rotating and fixed mirror is D then, the distance between the two points of reflection on the fixed mirror is given using the general equation for the perimeter of a circle $r\theta$:

$$S = 2D\delta\theta \tag{147}$$

Now consider a small wave packet or pulse of laser light. If the rotating mirror is rotating as the pulse strikes the rotating mirror (which

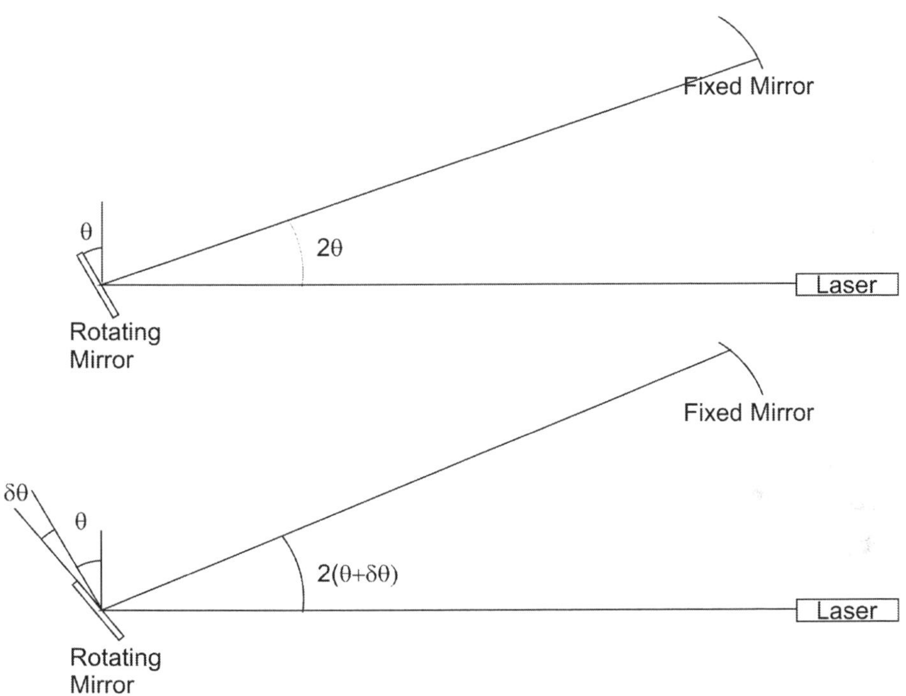

Figure 95: Reflection from the fixed mirror as the rotating mirror changes angle.

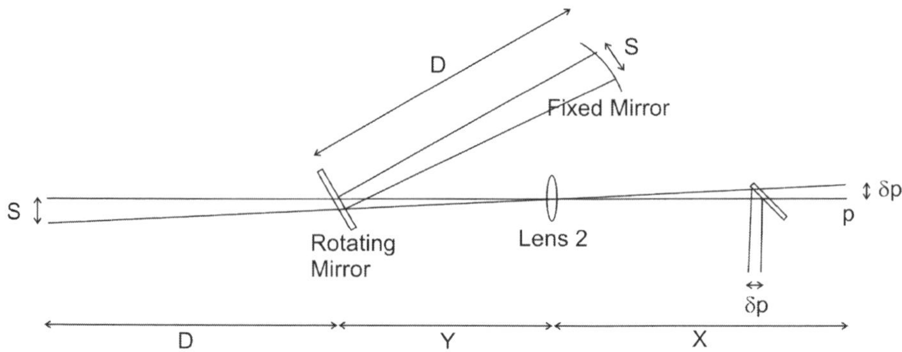

Figure 96: Considering the virtual image of the fixed mirror.

is at angle θ) the pulse will reflect up to the fixed mirror and then back to the rotating mirror. However the rotating mirror has rotated, by angle $\theta + \delta\theta$. If the mirror had not been rotating the returning pulse of light would have focused at the point p in Figure 94. Instead, it will focus at a different point p' which we must find.

It is simpler to consider the virtual image of the fixed mirror as shown in Figure 96. We now must consider (the thin) Lens 2. The virtual image of the two points of reflection on the fixed mirror are in the focal plane of Lens 2 but separated by a distance S, with one of the points on the principal axis. Thus, from thin lens theory we know that an object of height S in the focal plane will be focused in the plane of point p with a height $(-v/u)S$. In this equation v is the distance between the lens and image and u is the distance between the lens and the object. The negative sign can be ignored since we don't mind the image being inverted. The beam splitter directs a fraction of the light towards the microscope, giving a similar image with height δp. Therefore:

$$\delta p = \frac{v}{u} S = \frac{X}{Y + D} S \tag{148}$$

where X is the distance between Lens 1 and Lens 2 minus the focal length of Lens 1 and Y is the distance between Lens 2 and the rotating mirror. Substituting from Equation 147 gives:

$$\delta p = \frac{2DX\delta\theta}{Y+D} \tag{149}$$

The angle $\delta\theta$ depends on the rotational speed of the rotating mirror and the time it takes the light pulse to travel from the rotating mirror, up to the fixed mirror and back again (a distance $2D$). The time taken is:

$$t = \frac{2D}{c} \tag{150}$$

and recalling the rotation analogue of distance equals speed multiplied by time is $\theta = \omega t$ gives:

$$\delta\theta = \frac{2D\omega}{c} \tag{151}$$

where ω is the angular velocity of the rotating mirror and c is the speed of light. Substituting this into Equation 149 gives:

$$\delta p = \frac{4XD^2\omega}{c(Y+D)} \tag{152}$$

Rearranging gives our final equation for c:

$$c = \frac{4XD^2\omega}{\delta p(Y+D)} \tag{153}$$

We derived Equation 153 by assuming the image was created by a pulse of laser light at a particular position of the rotating mirror. However, Equations 147 and 151 only depend on the difference in the angle, that is the difference in the position of the rotating mirror in the time it takes light to travel between the mirrors. A continuous laser beam can be considered as a series of short pulses: the image of each pulse will be displaced by the same amount giving a single image. By making measurements of the displacement of the image, the angular velocity of the mirror and the necessary distances a value for the speed of light can be calculated.

Care should be taken with the laser used in this experiment. Refer to Section 2.7. You should not look into the microscope when the rotating mirror is stationary unless an attenuator is in place.

To begin, the equipment must be aligned. Turn on the laser. The lenses and beam splitter should already be fixed in the correct position. Using the edge of a piece of paper carefully follow the path of the laser beam to the rotating mirror. The rotating mirror should be positioned by hand so that the laser beam is reflected up to the fixed mirror.

The fixed mirror should the be adjusted so that the reflection of the laser beam strikes the center of the rotating mirror. After placing the attenuator in front of the microscope, the microscope should be focused on the image of the beam. It should appear as a small disk shaped dot. The microscope should be moved to bring the image into the center of the cross hairs and adjusted so that it is in focus.

The position of the microscope should be recorded using the Vernier scale attached to it. The microscope should be moved and then brought back to align with the image at the center of the cross hairs. The position should be recoded again. Repeating this around ten times allows the standard error on a single reading to be calculated.

The rotating mirror should then be turned on. It should rotate in a clockwise direction (looking down on the equipment as in Figure 94). The attenuator can be safely be removed. The angular velocity of the mirror is displayed, but sometimes takes a number of minutes to settle to a constant value. After waiting a while, the stability can be checked by recording ω every 10 seconds for a few minutes. This enables the calculation of a value for the standard error in ω.

Take the first reading by turning the rotation speed of the mirror up to a maximum and then once the speed has stabilised, bring the microscope image to the centre of the cross-hairs. Record the rotation speed of the mirror and the position of the traveling microscope using the Vernier scale. This should be repeated for a number of lower rotational speeds.

Measurements should also be taken with the mirror rotating anticlockwise. Finally the distances X, Y and D should be measured.

A graph should be plotted of microscope position δp against mirror rotational speed ω. The gradient of this graph is:

$$m = \frac{\omega}{\delta p} \tag{154}$$

The value of the gradient found can be substituted into Equation 153 to give a value for c. The error in a single measurement of δp and ω is found by calculating the standard deviation of the sets of 10 results taken earlier. Error bars should be added to the graph of microscope position against mirror rotational speed. These should be used to draw a maximum and minimum possible gradient and calculate the error in the gradient.

A suitable expression can be derived based on Equation 153 for the error in c. Hint, it may be useful to express $Y + D$ as another quantity W and find the error in W. Making this substitution into Equation 153 gives:

$$c = \frac{4XD^2}{mW} \tag{155}$$

It is interesting to consider which term contributes most to the error and is this a single dominant term? How could the random error in these measurements be reduced? Were there any sources of systematic error and can corrections be made? What effect does the beam splitter have, how big is it and is it significant?

6.2 Example Reports

The following section gives three examples of formal laboratory reports, based around different experiments to the examples already studied. The first example looks at finding the charge mass ratio of an electron, the second looks at radioactivity and the third looks at measuring the latent heat of vapourisation of liquid nitrogen. Although all these reports were graded as firsts, they are not reproduced here because they are perfect and on no account should they be copied. With careful thought there will always be ways in which they can be improved in terms of clarity, detail and content. However, they are provided here so the reader can see examples of the length, depth and quality that is required in a good report at undergraduate level.

6.2.1 Example Report: Charge Mass Ratio of an Electron

Abstract

Electrons emitted from an indirectly heated cathode were accelerated through known electric and magnetic fields and by varying the solenoid current, focussed into a narrow beam that then hit a fluorescent screen. When the beam was focused on the screen the electrons performed an integer number of helix rotations. For a series of anode – cathode potential differences, measurements of the solenoid current and number of helix rotations were taken. After careful analysis of the experimental data acquired, the charge:mass ratio, $\frac{e}{m}$, of the electron was calculated to be $2.99 \times 10^{11}\,\mathrm{Ckg}^{-1} \pm 0.55 \times 10^{11}\,\mathrm{Ckg}^{-1}$.

Introduction

The calculation of the charge:mass ratio of the electron was important in the early investigation and discovery of the structure of the atom and in proving the existence of units of matter smaller than the atom.

J.J. Thompson first carried out an experiment similar to this in 1897. He was attempting to disprove the widely believed idea that the hydrogen atom was the lightest particle in existence. His experiment, in which he measured the deflection of charged particles in a magnetic field perpendicular to the direction of motion of the particles, allowed him to show that the electron was nearly 2000 times lighter than the hydrogen ion. This experiment is based on an adaptation of Thompson's, first carried out by H. Busch in 1922.

This experiment aims to calculate $\frac{e}{m}$ and to test Busch's adaptation of Thompson's experiment.

Theory

The equation $F = Ee + Bev$ gives the force (measured in Newtons) on an electron of charge e (measured in Coulombs) and mass m (measured in kg), moving with velocity v (measured in m/s), in a magnetic field B (measured in Tesla) and electric field E (measured in volts/metre).

Since the acceleration due to this force is $a = \frac{(E+Bv)e}{m}$, neither e nor m can be found individually from this kind of experiment as only their ratio appears in the equation of motion. The ratio can be found from the paths taken by electrons passing through known electric and magnetic fields. Both electric and magnetic fields are required since the path in an electric field alone is independent of $\frac{e}{m}$. (In the time, $\delta t = \frac{\delta l}{v}$, that the electron takes to travel a short distance, δl, the direction of motion changes by $\delta\theta = \alpha\frac{\delta t}{v} = \alpha\frac{\delta l}{\delta v^2}$, where α is the acceleration transverse to the path. If the electron moves in only an electric field, α and v^2 are always proportional to $\frac{e}{m}$, therefore and the paths taken are independent of $\frac{e}{m}$. When there is also a magnetic field, α has a component proportional to v so the path depends on $(e/m)^{1/2}$, which can then be calculated.)

The most obvious way to perform this experiment, given the cathode ray tube, is to apply the magnetic field transverse to the electron beam and to calculate $\frac{e}{m}$ from the deflection of the beam across the screen. Using this method, it is very hard to achieve a good accuracy since the position of the spot is badly defined. These problems can be avoided by using the focusing effect of a longitudinal magnetic field on a nearly parallel beam of electrons. Since all the electrons started from the cathode with zero velocity (kinetic energies from thermionic emissions are usually less than a few tenths of an eV and can safely be ignored) and have been accelerated through a potential difference (the potential difference of the anode with respect to the cathode), as the beam of electrons emerges from the hole in the last disk of the anode, all the electrons have the same velocity, v, given by $eV = 1/2mv^2$. But, since the cathode has a finite area the directions of motion of the electrons are spread out i.e. the electron has a small component of velocity, v_t, perpendicular to the axis. In a uniform magnetic field parallel to the axis, the force experienced by the electron is perpendicular to the axis and to v_t and has magnitude Bev_t. This has the effect of rotating the direction of v_t but since it is always normal to it, to leave its magnitude unchanged.

The angular velocity of v_t is equal to the acceleration perpendicular to v_t divided by v_t i.e.

$$\omega = \frac{Bev_t/m}{v_t} = \frac{Be}{m}$$

ω is constant for a given B. This causes the electron to travel in a helical path towards the screen. The radius of this helix is

$$r = \frac{v_t}{\omega} = \frac{mv_t}{Be}$$

and the period of the motion is $T = (2\pi m)/Be$. The period depends upon e and m and not v, the velocity of the electron. Since the period is the same for all electrons, B can be chosen to give an integer number, n, of helical rotations as the electrons travel between the anode and the screen. This means that the electrons produce a sharply focussed spot on the screen – see Diagram (1). When n is not an integer number, the spot is badly focussed – see Diagram (2).

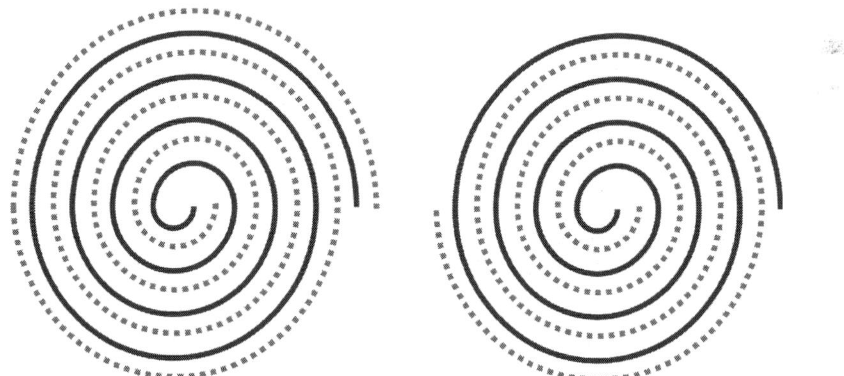

Left: Diagram (1) Electron paths when both have an integer number of helical rotations producing a focused spot. Right: Diagram (2) Electron paths when both do not have an integer number of helical rotations producing an unfocused spot.

If the anode-screen distance is L then the time taken for the electrons to travel this distance is L/v , then

$$n \approx \frac{L}{Tv} = \frac{LBe}{2\pi m}\sqrt{\frac{m}{2eV}}$$

So:

$$\frac{e}{m} = \frac{8\pi^2 n^2 V}{L^2 B^2}$$

Where $\frac{e}{m}$ is the charge:mass ratio of the electron, n is the number of helix rotations, V is the anode-cathode potential difference, L is the anode-screen distance and B is the magnetic field.

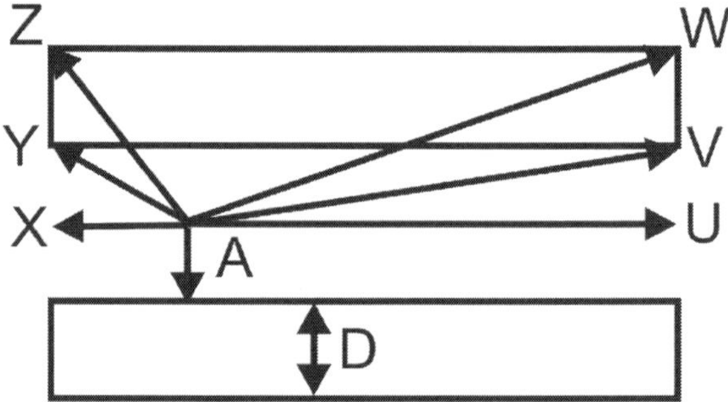

Diagram (3): Dimensions of a solenoid used to find the correction factor, K.

The field inside an infinitely long solenoid is given by $B = \mu_0 NI$ where μ_0 is the permeability of free space, N is the number of turns per metre and I is the current flowing in the solenoid. A correction factor of K is needed since the solenoid is only of a finite length. Where

$$K = \frac{X}{2D}\ln\left[\frac{A+D+Z}{A+Y}\right] + \frac{U}{2D}\ln\left[\frac{A+D+W}{A+V}\right]$$

and A, D, X, Y, Z, U, V, W are indicated on Diagram (3).

Experimental Methods

This experiment was carried out by finding the various fields for which the beam was focussed. For various values of V (the anode-cathode potential difference), readings of solenoid current were taken, using an ammeter, for points when the spot was in focus (i.e. integer values of n). As the current was increased from zero, the first time the spot was in focus was assumed to be $n = 1$, with n increasing by one each time the spot was in focus.

The solenoid was measured to find its dimensions to get a value for K and to calculate $\frac{e}{m}$.

After B had been calculated for each current, a graph of B^2 against n^2 was plotted so as the gradient was

$$\frac{e}{m}\frac{L^2}{8\pi^2 V}$$

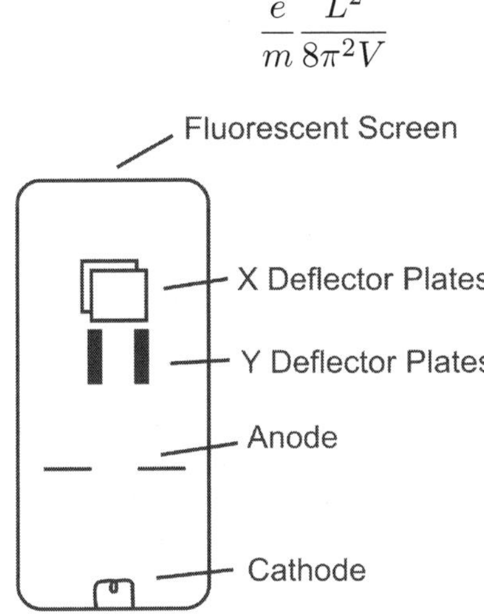

Diagram (4): Diagram of the cathode ray tube.

Diagram (4) shows the centre of the apparatus, including the anode, cathode and the fluorescent screen. This is surrounded by the solenoid.

Results

The value of K was calculated to be 0.87. L, the anode-screen distance was measured to be 0.051 m, the total length of the solenoid, L_0, was measured to be 0.13 m and the total number of turns on the solenoid was 1500. This gave N as 11500 turns/metre.

The gradients of the graphs gave $\frac{e}{m}$ for each (electric) potential difference as:

Voltage (V)	e/m (Ckg^{-1})
514	3.03×10^{11}
600	3.13×10^{11}
700	2.83×10^{11}
800	2.91×10^{11}
900	3.00×10^{11}

This gives an average value of $\frac{e}{m}$ as 2.99×10^{11} Ckg^{-1}.

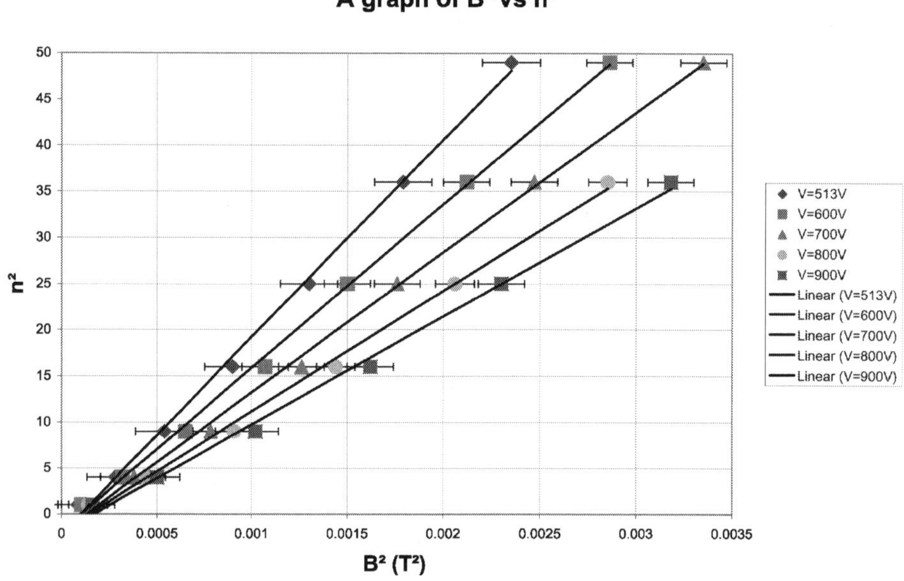

Diagram (5): Graph of B^2 versus n^2.

Discussion

After plotting the graphs for each of the (electric) potential differences, their gradients were found not to agree with each other exactly, but to vary slightly. An average value for the gradient was taken as the final answer that was around 1.7 times the accepted value. This means that the gradients of the graphs I plotted from my experiment were a little too large.

The error bars on the graph show the errors with respect to the uncertainty in the solenoid current. The error range was obtained by using the extremes of these error bars to calculate the gradient and then $\frac{e}{m}$. They do not show, what is probably the most significant source of error in this experiment, the error with respect to deciding the exact time when the spot was focussed on the screen. This was a visual measurement which is prone to error, which cannot be measured accurately.

Even though the gradients are a little large, the line of best fit still passes very close, but slightly to the positive side of the origin in all cases. This suggests that there could be some systematic error involved in these readings – which has a greater effect at larger values of n. Supporting this idea is the fact that the values of B should be multiples of the value for $n = 1$. This does not quite fit the experimental data. The values fall short of what this predicts them to be – and increasingly more so as n increases.

One possible cause of this could be the partial focussing of the beam of electrons before it emerges from the anode. To eliminate this the deflection of the spot when the electrons acquire a large transverse component of velocity in a known part of their path can be measured. This transverse component may be produced by applying a potential difference between one pair of deflector plates. The spot comes into focus when the electrons have traveled an integer number of turns in their helical path from the centre of the deflector plates to the screen. The distance L, in the earlier equation, can be substituted for the distance between the deflector plates and the screen, and $\frac{e}{m}$ calculated using the same methods.

Also, using an AC (electric) potential difference between the deflector plates could make it easier to see when the spot is in focus since the spot is then drawn out into a line and a collapse of the line into a spot indicates an integer number of helical rotations. Since the points lie so close to the lines of best fit, there is little evidence to suggest any significant random error in this experiment. As well as the extension involving applying AC (electric) potential differences to the deflector plates, a larger CRT could be used – as this would make it easier to see the spot and when it's in focus/not in focus.

Conclusions

From my experiment the charge:mass ratio of the electron is $2.99 \pm 0.55 \times 10^{11}\,\mathrm{Ckg}^{-1}$. Considering the inaccuracy in the visual "measurement" of when the spot is in focus, I consider my answer to be consistent with others' results, even though the accepted value of $1.76 \times 10^{11}\,\mathrm{Ckg}^{-1}$ is outside my error range, it is of the same order of magnitude and only a factor of 1.4 out. I consider Busch's adaptation of Thompson's experiment to valid – especially if the above improvement of applying an AC (electric) potential difference to the deflector plates behaves as predicted.

References

[1] Haliday, Resnick, Krane, "Physics", Fourth Edition, 1992, pp738–740

Appendix

V=513 V							
n	I1(A)	I2(A)	I3(A)	Iavg(A)	B(Tesla)	$B^2(T^2)$	n^2
1	0.72	0.73	0.74	0.73	0.00921	0.0000848	1
2	1.33	1.31	1.35	1.33	0.0168	0.000281	4
3	1.84	1.84	1.84	1.84	0.0232	0.000539	9
4	2.38	2.37	2.39	2.38	0.0300	0.000901	16
5	2.90	2.84	2.83	2.86	0.0361	0.00130	25
6	3.36	3.36	3.33	3.35	0.0423	0.00179	36
7	3.82	3.84	3.85	3.84	0.0484	0.00235	49

V=600 V							
n	I1(A)	I2(A)	I3(A)	Iavg(A)	B(Tesla)	$B^2(T^2)$	n^2
1	0.78	0.80	0.81	0.80	0.0101	0.000102	1
2	1.42	1.42	1.43	1.42	0.0179	0.000321	4
3	2.03	2.01	2.03	2.02	0.0255	0.000649	9
4	2.57	2.60	2.60	2.59	0.0327	0.00107	16
5	3.06	3.08	3.06	3.07	0.0387	0.00150	25
6	3.63	3.66	3.65	3.65	0.0460	0.00212	36
7	4.23	4.22	4.27	4.24	0.0535	0.00286	49

V=700 V							
n	I1(A)	I2(A)	I3(A)	Iavg(A)	B(Tesla)	$B^2(T^2)$	n^2
1	0.87	0.87	0.87	0.87	0.0110	0.000120	1
2	1.54	1.56	1.56	1.55	0.0196	0.000382	4
3	2.21	2.22	2.24	2.22	0.0280	0.000784	9
4	2.84	2.79	2.79	2.81	0.0354	0.00126	16
5	3.31	3.32	3.35	3.33	0.0420	0.00176	25
6	3.95	3.94	3.94	3.94	0.0497	0.00247	36
7	4.60	4.59	4.57	4.59	0.0579	0.00335	49

V=800 V							
n	I1(A)	I2(A)	I3(A)	Iavg(A)	B(Tesla)	$B^2(T^2)$	n^2
1	0.93	0.93	0.94	0.93	0.0117	0.000138	1
2	1.66	1.67	1.67	1.67	0.0211	0.000444	4
3	2.38	2.41	2.39	2.39	0.0301	0.000909	9
4	3.00	3.03	2.99	3.01	0.0380	0.00144	16
5	3.61	3.57	3.61	3.60	0.0454	0.00206	25
6	4.20	4.24	4.26	4.23	0.0534	0.00285	36

V=900 V							
n	I1(A)	I2(A)	I3(A)	Iavg(A)	B(Tesla)	$B^2(T^2)$	n^2
1	1.00	1.00	1.00	1.00	0.0126	0.000159	1
2	1.77	1.78	1.77	1.77	0.0223	0.000499	4
3	2.56	2.52	2.52	2.53	0.0319	0.00102	9
4	3.18	3.17	3.22	3.19	0.0402	0.00162	16
5	3.85	3.78	3.78	3.80	0.0479	0.00230	25
6	4.48	4.48	4.45	4.47	0.0564	0.00318	36

6.2.2 Example Report: Investigating Radioactivity

Abstract

Some of the properties of radioactive count rate and the Geiger Muller tube were investigated. The binomial distribution for independent events was approximated by the Normal and Poisson distribution. The Poisson was found to be a better approximation since the probability of decay was small. The dead time of the Geiger Muller tube was found to be 495 ± 73 μs. The inverse square law for the count rate was tested with a gamma and a beta source, the results showed an almost perfect fit to theory. Gamma rays were found, as expected, to be more penetrating than beta particles. Exponential attenuation was also investigated with lead plates. The theoretical equation $M = M_0 e^{-\mu t}$ was found to fit the experimental data quite well, but a different equation, $M = M_0 e^{-\sqrt{k}t}$, was found to provide a much improved fit, especially where the standard thickness lead plates were not combined to provide additional thicknesses to test.

1. Introduction

1.1 History of radioactivity

1.1.1 Antoine Henri Becquerel (1852 – 1908)

In March of 1896, a French physicist, Antoine Henri Becquerel, placed some wrapped photographic plates into a draw along with some crystals containing uranium. Much to his surprise, when he returned to the draw sometime later he discovered the plates were exposed. He concluded that the uranium crystals must have been emitting something, since the exposure had occurred without the presence of the usual initiating energy source – sunlight. Becquerel, though, did not pursue his discovery of radioactivity any further. This was left to the Curies.

1.1.2 Marie (1867 – 1934) and Pierre (1859 – 1906) Curie

Becquerel had already discovered that the 'ray' emitted by uranium could turn air into a electrical conductor. The Curies followed this up, testing various materials. On February 17th 1898 they discovered that an ore of uranium, pitchblende, produced a current over 300 times more than that produced by pure uranium. After repeating the experiment and obtaining the same answer, they concluded that some very active substance other than uranium must exist within the pitchblende. In a paper detailing their experiment, they hypothesised the existence of a new element that they named polonium, and coined the phrase 'radioactive' to describe elements like uranium and polonium. Continuing their work the Curies discovered another radioactive element, radium. Marie continued her work after Pierre's death and discovered there was a decrease in the rate of radioactive emissions over time and that this decrease could be calculated and predicted.

In 1903, Becquerel and the Curies together received the Nobel Prize in physics. This award was for their discovery of radioactivity and their other contributions in this area.

1.1.3 Ernest Rutherford (1871 – 1937)

Up until Ernest Rutherford discovered the structure of the atom in 1911, there was very little scientific explanation for what caused radioactivity and how the emissions affected the original atoms. In a series of experiments he bombarded a piece of gold foil with alpha particles emitted by a radioactive material. Most of the particles passed through the foil undisturbed, suggesting that the foil was made up mostly of empty space rather than of a sheet of solid atoms. Some alpha particles, however, bounced back, indicating the presence of solid matter. Rutherford concluded that atoms consisted primarily of empty space surrounding a well-defined nucleus.

Many more years passed until the full effect of radioactive decay, on the atoms undergoing the decay, was fully understood.

1.2 Quantitatively measuring the rate of radioactive decay

The radioactivity of a substance is measured by how many decays occur in a unit time. Both Becquerel and the Curies have units of radioactivity named after them. The Curie (Ci) is defined as $1\text{Ci} = 3.7 \times 10^{10}\,\text{decays}\,\text{sec}^{-1}$ and the Becquerel (Bq) is defined as $1\,\text{Bq} = 1$ decay sec^{-1}.

1.3 Modern Day Uses for Radioactive Isotopes

Today, radioactive isotopes are used widely in industry for thickness control and other forms of testing; in health care as tracers, cancer treatments and in the sterilisation of surgical instruments; in domestic appliances such as smoke detectors; and for power production on spacecrafts, navy vessels and power stations.

1.4 Aims of this experiment

This experiment is designed to test the background statistical theory behind radioactive decay and investigate one set of apparatus used for measuring radioactivity – the Geiger Muller tube. It also seeks to investigate some of the properties of radioactivity such as exponential decay and the ability of certain materials to scatter/absorb particular types of radiation. These and other properties are made use of in many of the processes mentioned in section 1.3.

2 Theory

2.1 Binomial Distribution

Radioactive decay is a random process, so the probability of any particular atom decaying is independent of whether any other particular atom has decayed. This is characteristic of the Binomial distribution.

$$P(r) =^n C_r p^r q^{n-r}$$

If an atom has probability $\lambda \delta t$ of decaying in a short time interval δt, given it has not decayed already, the probability distribution function for survival is (for independent probabilities)

$$F(t + \delta t) = F(t).(1 - \lambda\delta t) : (1)$$

Where $F(t + \delta t)$ is the probability of survival at time $t + \delta t$, $F(t)$ is the probability of survival at time t and $(1 - \lambda\delta t)$ is the probability it doesn't decay in next δt. If a Taylor expansion is performed on the left hand side of equation (1) it becomes:

$$F(t) + \frac{dF}{dt}\delta t + ... = F(t).(1 - \lambda\delta t) : (2)$$

As $\delta t \to 0$, (2) becomes:

$$\frac{dF}{dt} = -\lambda F(t) : (3)$$

Equation (3) is a 1st order differential equation for $F(t)$, which has the solution:

$$F(t) = Ce^{-\lambda t} : (4)$$

Since $F(0) = 1$ then $C = 1$ i.e. the particle had not decayed at $t = 0$. Equation (4) can be multiplied by N_0, the total number of atoms present to give:

$$N = N_0 e^{-\lambda t} : (5)$$

This tells us the number of particles, N, left after time t.

2.1.1 Normal and Poisson Approximations to the Binomial

Approximations to the Binomial can be evaluated using the Normal or the Poisson distribution. The Normal approximation (when n and np are large) is:

$$P(r) =^n C_r p^r q^{n-r} \approx \frac{1}{\sqrt{2\pi npq}} e^{-(r-np)^2/2(npq)^2}$$

The Poisson approximation (when n large and p small) is:

$$P(r) = {}^{n}C_{r}p^{r}q^{n-r} \approx \frac{(np)^{r}}{r!}e^{-np}$$

2.2 Chi Squared Test

In order to see how well a theoretical distribution fits a series of experimental results a Chi Squared test can be performed. A good fit requires that $\chi^2 \approx v$ where v is the number of degrees of freedom.

$$\chi^2 = \sum \frac{(y_{\text{theoretical}} - y_{\text{experimental}})^2}{y_{\text{theoretical}}}$$

The reduced value of χ^2 which can be looked up in tables to give a probability is

$$\chi_v^2 = \frac{\chi^2}{v}$$

If this probability is greater than 0.05 the experimental data can be said to fit the theory.

2.3 The Geiger Muller Tube

In order to measure the decay rate of the radioactive sources a Geiger Muller Tube was used: see Figure 2.2.

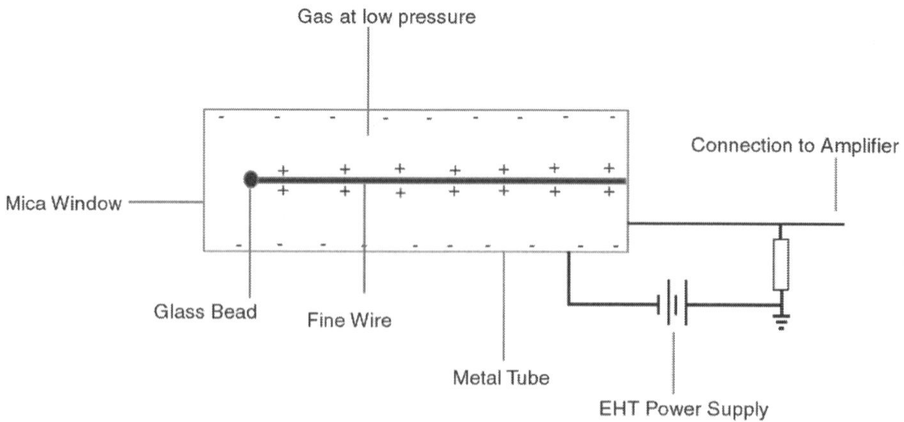

Diagram 2.2: The Geiger Muller Tube.

A large potential difference is applied by the EHT power supply, so as the tube becomes negatively charged with respect to the wire. This (electric) potential difference is decided with reference to a number of conditions. It needs to be above the threshold frequency, below the (electric) potential difference that causes electrical discharge through the gas and on the plateau of the calibration graph. When any form of radiation enters the tube thorough the thin mica window it ionises a gas atom. The electron created moves towards the central wire and the gas ion moves towards the tube (due to electro-static attraction). The ion moves more slowly than the electron since it is move massive i.e. by Newton's second law its acceleration is lower. Due to the high emf of the EHT supply, the electron is accelerated rapidly and causes the ionisation of another atom. This builds rapidly into an avalanche of electrons. The velocity of propagation of the electrons is high enough so that as a result of the avalanche of electrons the charge available for collection by the wire has a constant value, irrespective of where in the tube the original ionisation occurred.

When the electrons from the avalanche have been collected the slowly moving sheath of positive ions acts as a partial electrostatic shield and therefore reduces the field below that necessary for ionisation by collision. This avalanche of electrons has the effect of producing a large current in the external resistance. This can then be used to increment a counter for each particle/ray reaching the tube. The tube is now ready to make another count.

The sheath of positive ions can eject electrons from the tube, leading to more avalanches and multiple counts for only one decay. Alcohol is usually added to the gas in the tube to quench this extra discharge. The alcohol does this since it has a lower ionisation energy than the gas, and so is ionised first. The energy that was used to release extra electrons is now used in dissociating the alcohol molecule.

2.4 Dead Time

Geiger Muller tubes have limitations. One important limitation in this experiment is dead time. This is caused as an after effect of the electrical breakdown of the gas filling the Geiger Muller tube. Since the tube will not register any other counts for a short period of time after each count, a significant proportion of the overall count can be lost.

If two identical sources are used and N_1 is the number of counts with only source 1 present, N_2 is the number of counts with only source 2 present and N_3 is the number of counts with both source 1 and 2 present then the dead time can be shown to be:

$$t = \frac{1}{N_3} - \sqrt{\frac{1}{N_3^2} - \frac{N_2 + N_1 - N_3}{N_1 N_2 N_3}}$$

Proof of this equation is given in Appendix 8.1.

2.5 Isotopes and Types of Decay

Atoms consist of protons, neutrons and electrons. In order to have a zero charge an atom must have the same number of protons and electrons. Atoms of different elements have different numbers of protons. Some elements have more than one isotope - i.e. they have a different number of neutrons (but the same number of protons). Some isotopes are radioactive and decay, in order to become more stable, in one of the following ways.

Alpha Decay – The nucleus loses two neutrons and two protons.

Beta Minus Decay – A neutron (from the nucleus) decays to become a proton; and an electron and an anti-electronneutrino that are both emitted.

Beta Plus Decay – A proton (from the nucleus) decays to become a neutron; and a positron and an electron neutrino that are both emitted.

Gamma Rays – Can be created in a number of ways as the atom decays, but the method relevant to the sources used in this experiment is the annihilation of an electron from the air and a positron emitted during beta plus decay.

2.6 Shape of Calibration Graph

The counting starts at a threshold (electric) potential difference indicated by A on Diagram 2.6. This is when the field becomes strong enough so that ionisation of the gas atoms occurs. The count rate then rises rapidly up to the plateau beginning at B. The Geiger Muller tube is operated on the plateau at the (electric) potential difference B, since any slight variation in the (electric) potential difference produced by the (electric) potential difference supply will then have a negligible effect on the recorded counting rate. The count rate rises rapidly at higher (electric) potential differences since breakdown has occurred i.e. the quenching of the alcohol is not enough to stop multiple counts for a single decay. Ideally the plateau should be flat, but in practise it slopes with a small positive gradient since as the (electric) potential difference increases the probability of spurious counts also increases.

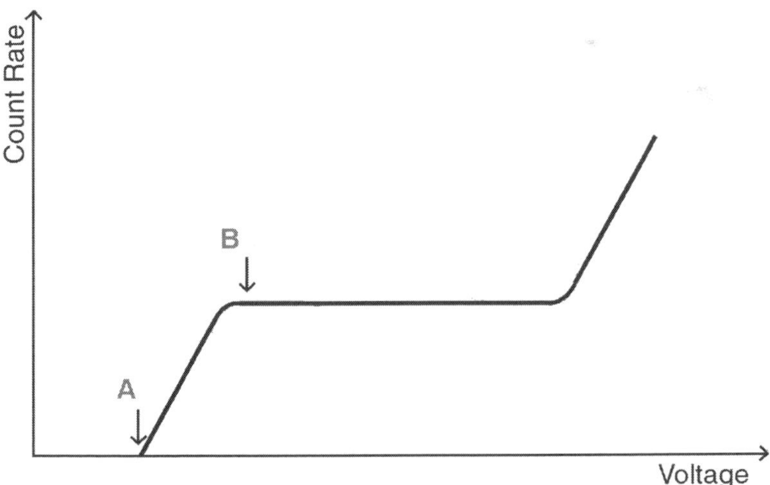

Diagram 2.6: Idealalised Calibration Graph

3. Experimental Methods

3.1 Sources Used

All these sources have very long half lives so as the count rate will be effectively constant over the whole period of the experiment.

Source	Radiation Emitted	Decay Equation	Half Life
^{90}Sr	β^-	$^{90}_{38}\mathrm{Sr}\rightarrow^{90}_{39}\mathrm{Yr}+e^- + \overline{\nu}$	29.02 yr
^{22}Na	$\beta^+ \rightarrow \gamma$	$^{22}_{11}\mathrm{Na}\rightarrow^{22}_{10}\mathrm{Ne}+e^+ + \nu$	2.60 yr
		$e^- + e^+ \rightarrow 2\gamma$	
^{204}Tl	β^-	$^{204}_{81}\mathrm{Tl}\rightarrow^{204}_{82}\mathrm{Pb}+e^- + \overline{\nu}$	3.78 yr

3.2 Diagram of Apparatus

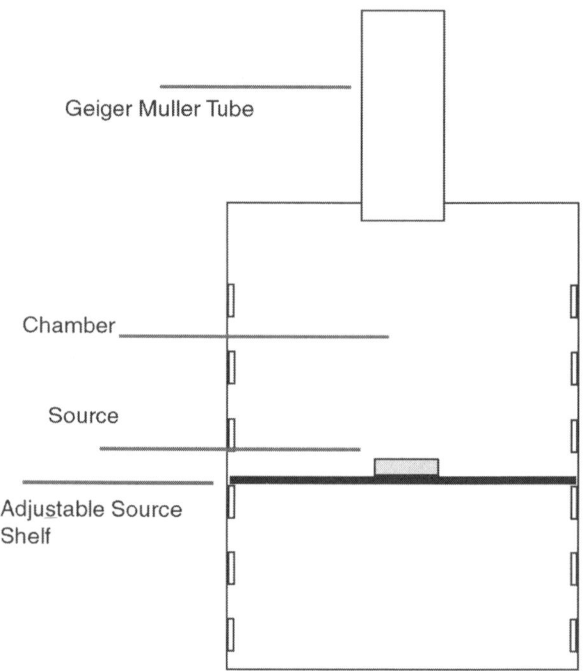

Diagram 3.2: The apparatus used

3.3 Finding the Optimal Voltage at which to run the Geiger Muller Tube

In order to determine the (electric) potential difference, B, at which to run this Geiger Muller tube, measurements were taken of the count rate of the ^{90}Sr source and EHT emf. The computer was set to take the number of counts over a 10 second period, at (electric) potential differences from 0 V to 1200 V in 10 V steps. The count rate was calculated by dividing the count by the number of seconds. Section 4.1

shows the graph obtained by plotting the (electric) potential difference against count rate.

3.4 Shape of the Normal Approximation

From Section 2.1.1 it is noticed that for a Normal approximation to a Binomial distribution to be accurate, n and np must be large. This can be achieved by measuring a large number, n, of count rates. Since np is the mean of the Normal distribution, the mean count rate should also be large – around 100 should be sufficient. The mean count rate can be varied by adjusting the position of the source in the chamber.

This time the computer was set to take 1000 readings of the count over 1 second (which gives the count rate). The mean and variance of these results was calculated using the standard formulae – see Lyons (1991). The theoretical Normal distribution with this mean and variance was calculated and then multiplied by 1000 so as the theoretical and experimental graphs could be plotted and compared on the same scale. Section 4.2 shows the graph obtained.

3.5 Shape of the Poisson Approximation

Again from Section 2.1.1 it is noticed that for a Poisson Approximation to a Binomial distribution to be accurate, n must be large and p must be small. In order to ensure p was small, a scatter/absorber material was inserted between the source and the tube so as the mean was reduced to around $2-4$ counts/second. Again, the computer was set to take 1000 readings of the count rate over 1 second. The mean of these readings was calculated. The theoretical Poisson distribution with this mean was then calculated and multiplied by 1000 so as the theoretical and experimental graphs could be plotted and compared on the same scale. Section 4.3 shows the graph obtained.

3.6 Chi Squared Test

For the Normal and Poisson approximation the values of χ^2 and prob(χ^2)

were calculated. This enables the closeness of the fit to be analysed quantitatively.

3.7 Background Count

The background count (i.e. the count rate with no source present) was measured and subtracted from all subsequent measurements. The count rate was found to be 0.48727 counts $\sec^{-1}$.

3.8 Dead Time Measurement

In order to calculate the dead time, t, values of N_1, N_2 and N_3 were obtained. A value for N_1 was obtained by performing a series of measurements of the count over a period of 10 s with just one ^{204}Tl source present. A value for N_2 was obtained by the same method with the other ^{204}Tl source present. Finally, a value for N_3 was obtained by the same method with both the sources present. Putting these values into equation 2.4 gives a value for the dead time. Section 4.4 gives the results and the final value for dead time.

3.9 Dead Time and the Background Count Rate

The background count was adjusted for the dead time. The adjusted background count rate was now found to be 0.48739 counts $\sec^{-1}$. This new value was subtracted from all subsequent results.

3.10 The Inverse Square Law

Since the particles emitted by a radioactive source are random in both time and direction, the count rate should fall with the square of the distance between the source and the Geiger Muller tube. The experiment was performed by placing the source on each of the shelves in the chamber and then taking a reading of the count rate over a 10 second period. The two sources tested were ^{90}Sr and ^{22}Na. The intershelf distance was constant and found to be 12.8 mm. Graphs of $1/\sqrt{M}$, where M is the count, against distance could then be plotted.

Using the intershelf distance method meant that there was no need to know the absolute distance between the source and the Geiger Muller tube window. In order to plot the results the distance of the first shelf was arbitrarily chosen to be 1m. This method still allowed for comparison of the results and for the inverse square law to be tested.

After the points were plotted and a line of best fit was added, Excel was used to give the equation of the line and also a value of R^2. From the equation of the line the x-axis intercept was also calculated. The results are given in section 4.5.

3.11 Exponential Attenuation

The absorption of particles by a certain material can be investigated by placing the source at a constant distance from the Geiger Muller tube and inserting various thicknesses of the material in between. Alpha and Beta particles lose energy progressively and have a definite range in a certain material but gamma rays are expected to be absorbed exponentially.

In order to test whether this was the case, the ^{22}Na source and various thicknesses of lead were used. These plates were inserted in various combinations in between the source and the Geiger Muller tube and the count rate measured. The results are provided in Appendix 8.2.

Thickness (inches)	Code	Density ($\mathrm{g\,cm^{-3}}$)	$\mathrm{mg\,cm^{-2}}$
0.250	t	11.71	7435
0.125	s	11.44	3632
0.064	r	11.63	1890
0.032	q	15.13	1230

From preliminary graphs it was concluded that the gamma were absorbed exponentially. Kaplan (1956) gives the equation $M = M_0 e^{-\mu t}$ where M is the count rate, M_0 is the count rate with no lead being used, t is the thickness of the material and μ is the linear attenuation coefficient. By comparing

$$\begin{aligned} \ln M &= \ln M_0 - \mu t \\ y &= c - mx \end{aligned}$$

it can be seen that if a graph of $\ln M$ against t is plotted with the intercept fixed at $\ln M_0$, the gradient of this graph would be the constant $-\mu$. Graphs are shown in Section 4.6

4 Observations and Results

4.1 Calibration Graph

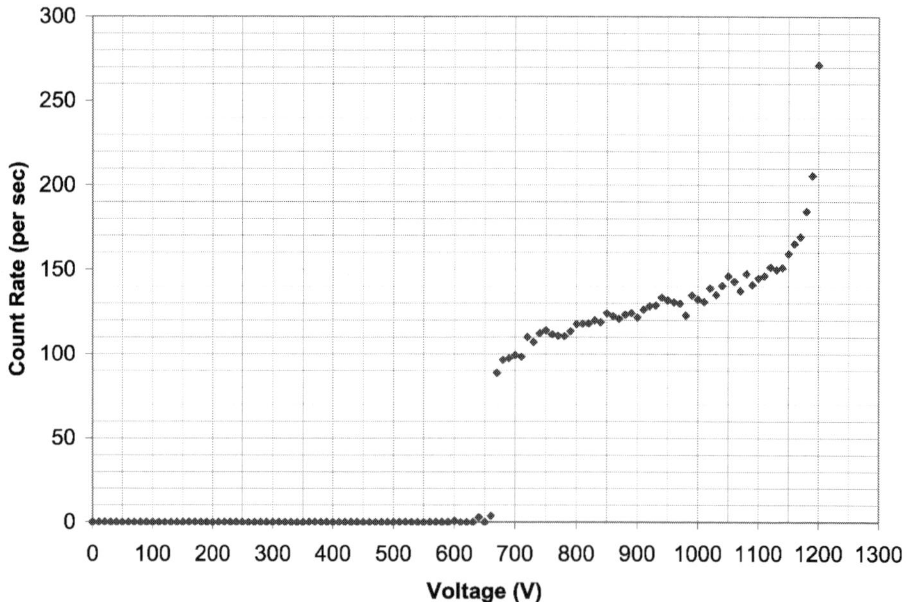

Graph 4.1: Calibration Graph

The threshold (electric) potential difference (A) can be seen to be 650 V and the start of the plateau (B) is at 750 V. For the rest of the experiment the Geiger Muller Tube was run with a emf supply of 750 V.

4.2 Graph of the Normal Approximation

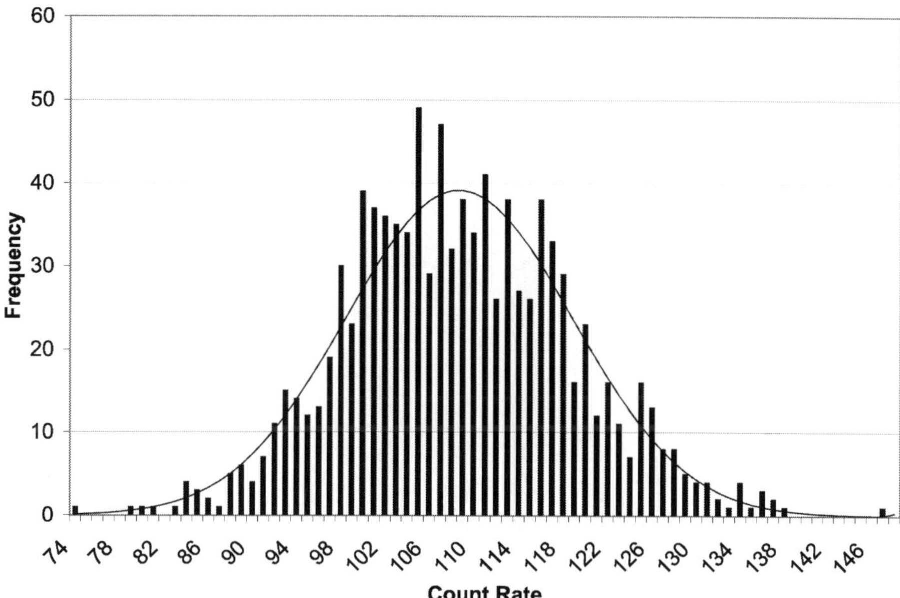

Graph 4.2: The Normal Approximation

The bars show the experimental results and the line shows the theoretical results. The mean is 109, the variance is 10.2, the prob(χ^2) is 0.47 and the χ^2 value is 73.1.

4.3 Graph of the Poisson Approximation

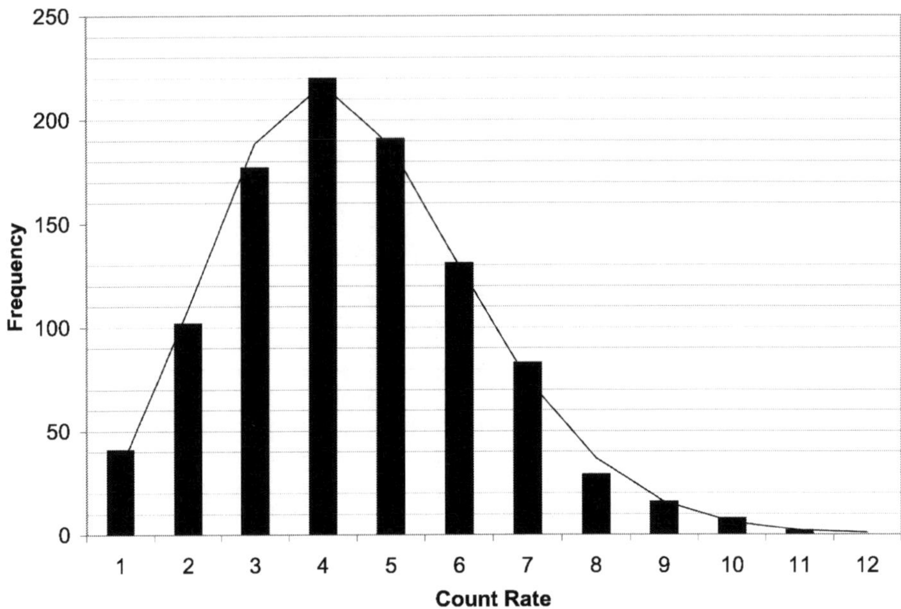

Graph 4.3: The Poisson Approximation

Again the bars show the experimental results and the line shows the theoretical results. In order to sufficiently reduce the mean count rate a piece of aluminium 0.032 inches thick was found to be necessary. The mean is 3.5, the variance is 3.5, the prob(χ^2) is 0.61 and the χ^2 value is 7.29.

N_1	N_2	N_3	Dead Time (10^{-6})s
362	281	542	591
335	298	510	765
320	300	548	424
316	300	541	450
307	279	567	115
347	274	513	690
303	265	544	156
341	300	542	572
369	267	491	964
347	300	563	464
331	282	541	437

The N_x values shown above are already corrected for the background count and then the dead time was calculated. The mean, standard deviation and standard error on the mean were calculated – see Lyons (1991). The mean is $510\,\mu$s and the standard error on the mean is $70\,\mu$s.

4.5 Inverse Square Law

4.5.1 Beta Source

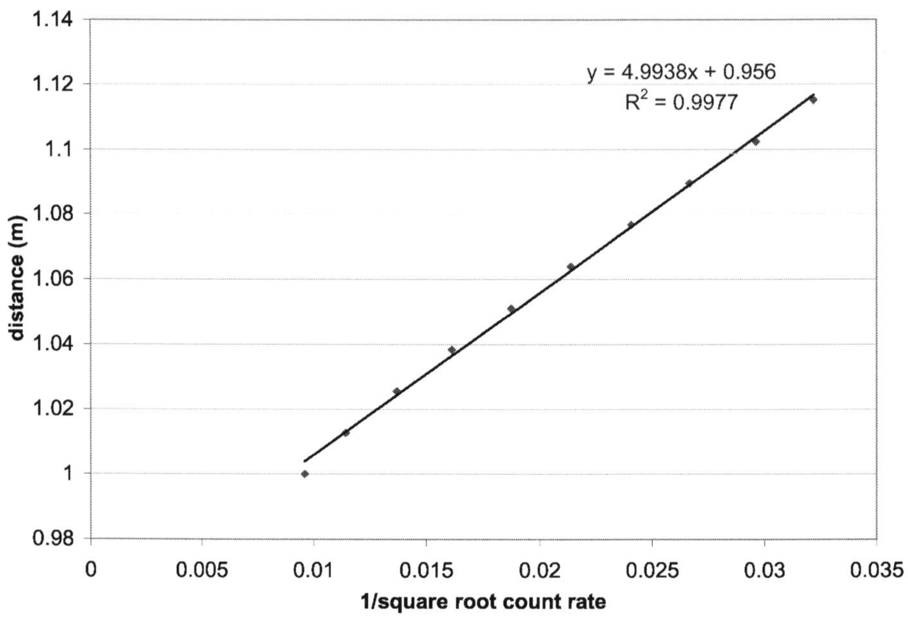

Graph 4.5.1: The Beta Source

With the beta source, the x-axis intercept was found to be at -0.19 and the R^2 value was found to be 0.9977.

4.5.2 Gamma Source

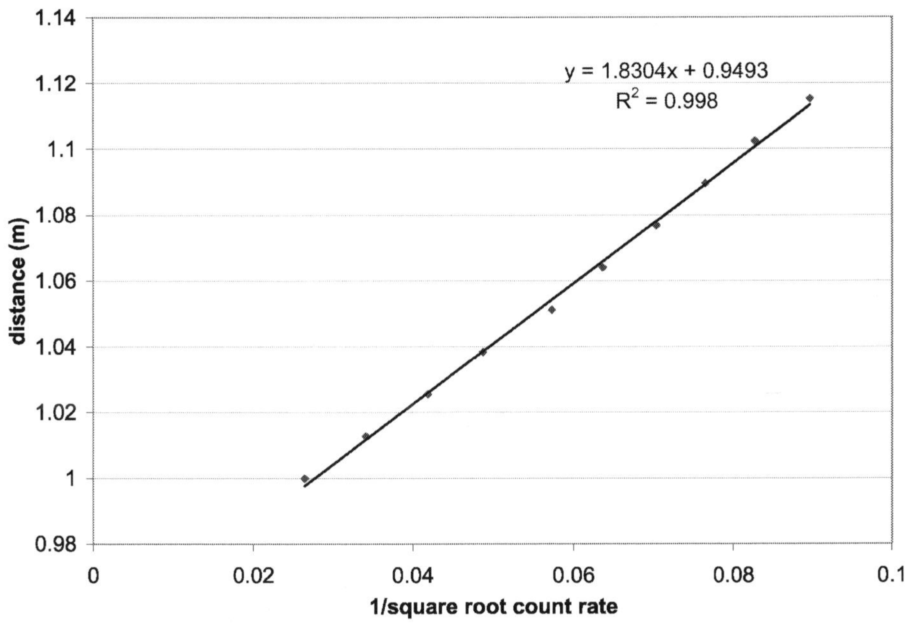

Graph 4.5.2: The Gamma Source

With the gamma source, the x-axis intercept was found to be at -0.52 and the R^2 value was found to be 0.998.

4.6 Exponential Attenuation

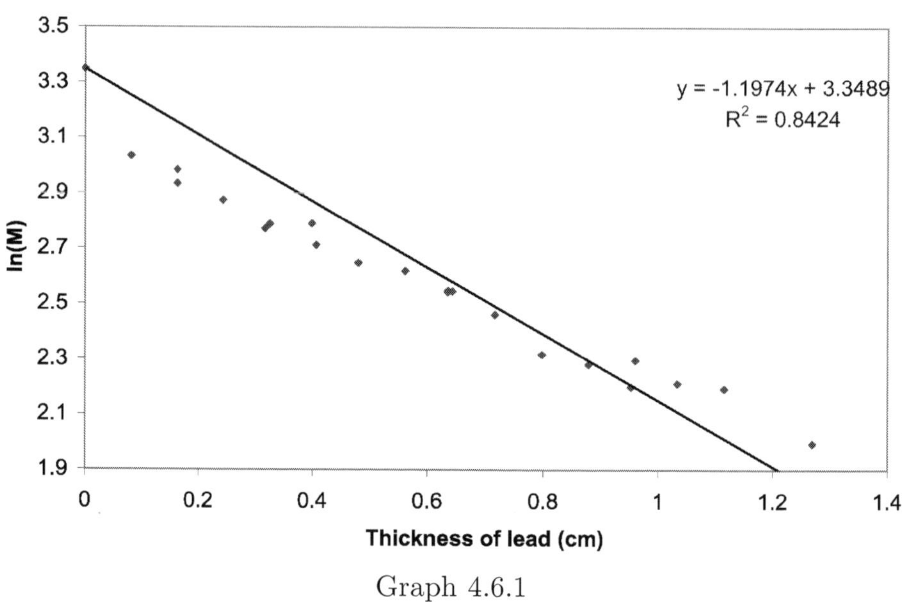

Graph 4.6.1

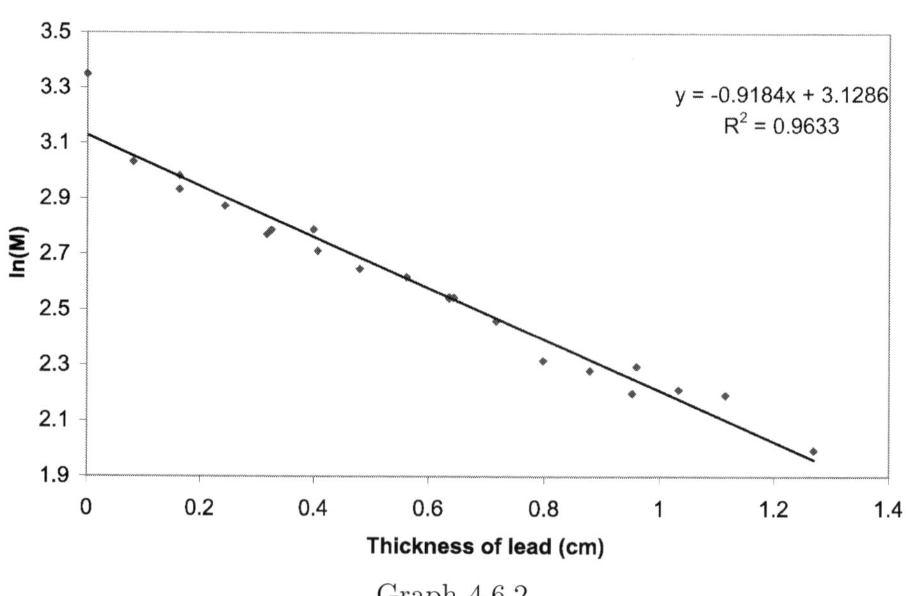

Graph 4.6.2

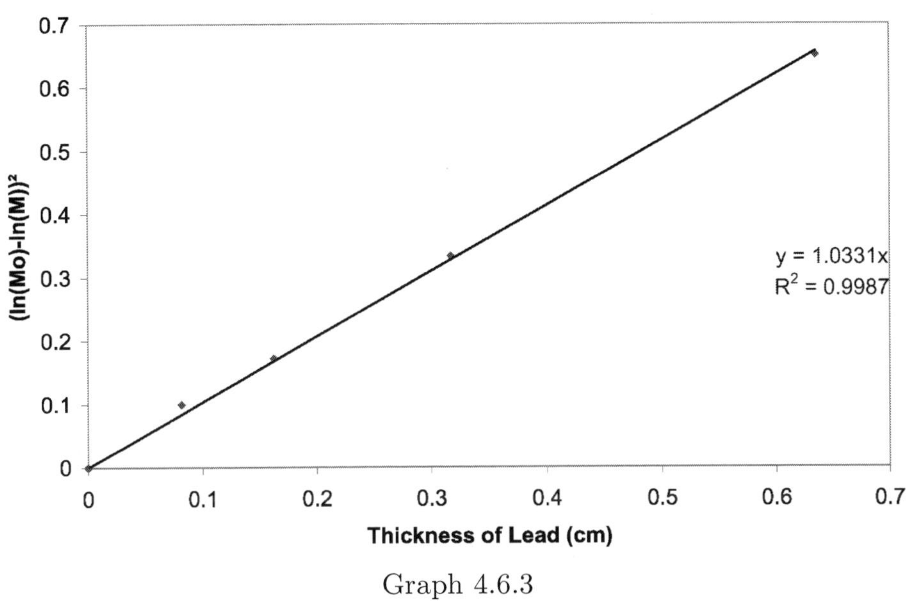

Graph 4.6.3

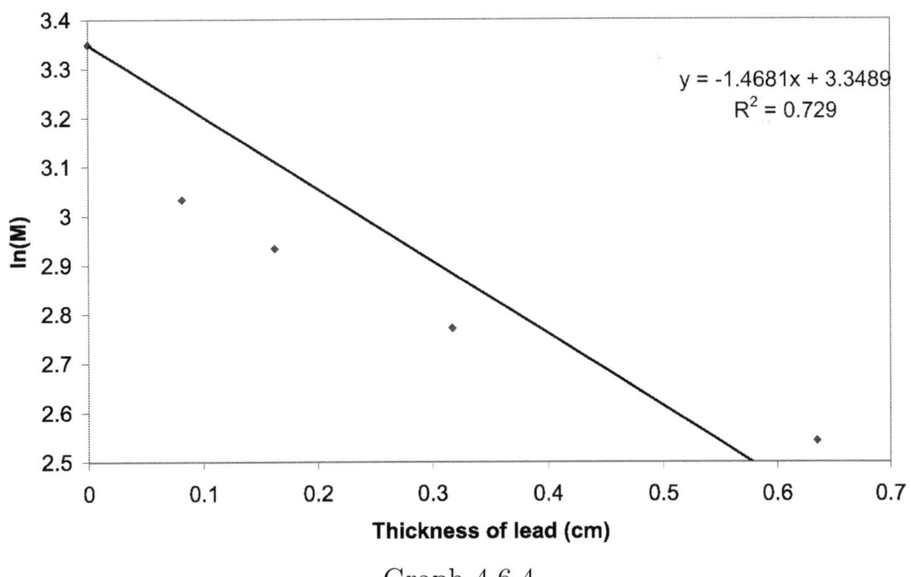

Graph 4.6.4

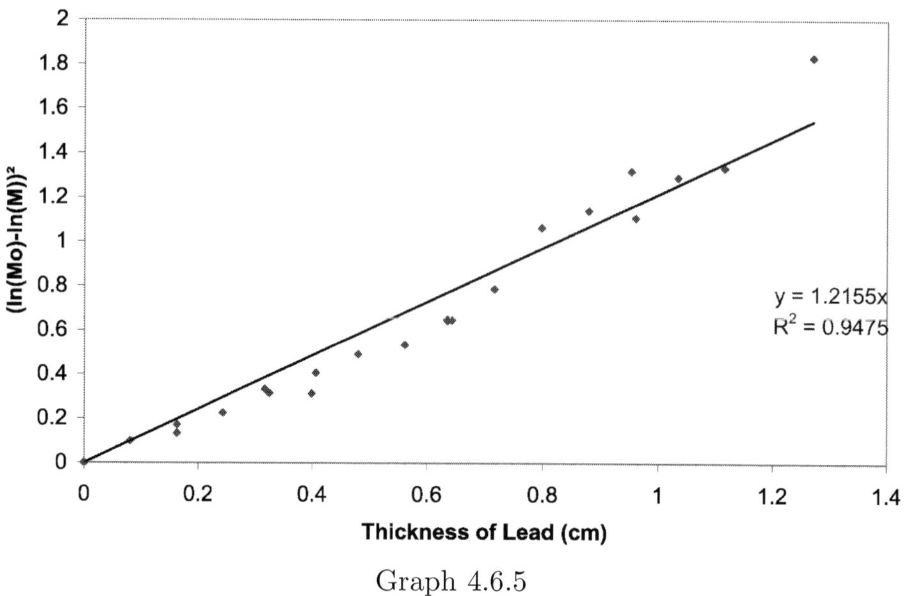

Graph 4.6.5

Using Graph 4.6.1 μ was calculated to be 1.1974. Using:

$$\frac{\mu}{\rho} = \frac{\sigma}{M_a}$$

from Kaplan (1956), where μ is the linear attenuation coefficient, ρ is the average density of the lead, σ is the cross sectional area of a scattering centre (a lead atom) and M_a is the mass of a lead atom. σ was found to be $3.33 \times 10^{-23}\,\text{cm}^2$. Since 1 barn$= 1 \times 10^{-24}\,\text{cm}^2$, $\sigma = 33.32\,\text{barns}$.

5 Discussion

The plateau on the calibration graph was found to start at around 750 V. All readings for the remainder of the experiment were carried out with the Geiger Muller tube running at this (electric) potential difference.

The count rate was found to fit the Poisson distribution more closely than the Normal. The prob(χ^2) value for the Poisson was 0.61 where as for the Normal it was only 0.47. Considering the conditions for both these approximations outlined in 2.1.1, the number of count rates measured should have been higher than 1000. This should ensure the conditions are satisfied more exactly and should increase the closeness of fit between the theoretical and experimental results.

The dead time was found to be $490 \pm 70\,\mu s$. This value is of the order of that expected for a typical Geiger Muller tube of the type used. Therefore this can be considered to be a reliable value. Perhaps, in order to increase the accuracy of this part of the experiment more readings could be taken, so as to decrease the error, and increase the certainty of the measurement.

Graphs 4.5.1 and 4.5.2 for the beta and gamma emitting sources respectively, very closely obey the inverse square law. This is shown by the closeness of the R^2 value to 1. The x axis intercept for the gamma source is lower than that for the beta source, suggesting that the gamma rays, as expected, are penetrating further into the Geiger Muller tube than the beta particles. This could indicate that the Geiger Muller tube was acting as a sensitive volume in the case of the gamma source and a sensitive area in the case of a beta source (beta particles being much less penetrative). There seemed to be little trouble with the beta particles not having enough energy to penetrate the mica window of the Geiger Muller tube since a significant count rate was always obtained at distances comparable with the gamma source.

Kaplan (1956) gave the equation for exponential attenuation as $M = M_0 e^{-\mu t}$ where M is the count rate, M_0 is the count rate with no lead being used, t is the thickness of the lead and μ is the linear attenuation coefficient. Looking at graph 4.6.1, where the intercept was fixed at $\ln M_0$ the data fits the equation very well. It provides an even better fit if the intercept is not fixed, as shown in graph 4.6.2. While taking the first few sets of data and plotting some preliminary graphs it did not look like the data was going to fit the equation from Kaplan (1956). Graph 4.6.4 shows the data for the first 4 combinations of the lead tested. (These were all single pieces of lead of the specified thickness. In all further trials, a combination of lead plates was used and the total thicknesses was simply taken to be the sum of the individual thicknesses.) This inaccuracy led a search for another equation that the data fitted more accurately. Using the relationship $M = M_0 e^{-k\sqrt{t}}$ and plotting $(\ln M_0 - \ln M)^2$ against t, with the same four results as before, there appeared to be an almost perfect straight line fit as graph 4.6.3 shows. Using the modified equation produced an R^2 value of 0.9987, this is much closer to 1 than the R^2 for the graph using the equation from Kaplan (1956) which was 0.729.

After performing further tests with other thicknesses of lead the whole set of results was plotted using each of the above equations. Graph 4.6.5 shows the whole set of data plotted with the modified equation and graph 4.6.1 shows the whole set of data plotted with the original equation from Kaplan (1956). The modified equation still produces a better fit with a R^2 value of 0.9475 compared to 0.8424. (The graph plotted for the modified equation, should go through the origin, so the line of best fit was fixed to do this. This graph for the equation from Kaplan (1956) should go through $\ln M_0$ so it was fixed to do this. Therefore a valid comparison can be drawn between these two graphs and not between graph 4.6.2 where the intercept was not fixed. However fixing intercepts potentially disguises any systematic error which is present in the results.)

Further research would be needed to determine why the modified equation is especially appropriate when single blocks of lead are used. It would be interesting to test with other thicknesses made up solely of one block of lead to see if the modified equation is indeed better or if the closeness of fit of the modified equation is an experimental one off. As the number of results increased the fit of the Kaplan (1956) equation improved and the fit of the modified equation deteriorates slightly, but in the limit of the results taken, still remains more accurate.

6 Conclusions

The start of the plateau for the Geiger Muller tube was found to be 750 V. Both the Normal and Poisson distribution are valid approximations to the Binomial in this case, although the Poisson provides the closer fit between experimental data and theory. The dead time of the Geiger Muller tube was calculated to be $490 \pm 70\,\mu$s. The count rate for both the gamma and beta sources was found to accurately obey the inverse square law. The data gathered while investigating exponential attenuation supports the equation, $M = M_0 e^{-\mu t}$ given in Kaplan (1956), but the modified equation, $M = M_0 e^{-k\sqrt{t}}$ fits the data more accurately. The cross section of a scattering centre was found to be 33.32 barns.

7 References

Taylor, Denis (1957) The Measurement of Radio Isotopes, Methuen + Co.

Burcham (1963) Nuclear Physics: An Introduction, McGrawhill Book Company

Kaplan (1956) Nuclear Physics, Addison-Westley

Halliday, Resnick, Krane (1992) Physics, Wiley (Chichester)

Boas, Mary (1983) Mathematical Methods in the Physical Sciences, Wiley (Chichester)

Lyons, Louis (1991) A Practical Guide to Data Analysis for Physical Science Students, Cambridge University Press (Cambridge)

Barford, W. C. (1985) Experimental Measurements: Precision, Error and Truth, 2nd Edition, Wiley (Chichester)

8 Appendix

8.1 Dead Time

If N counts are recorded in 1 second, and the tube is dead for a time t after each count then in 1 second the tube will be dead for Nt seconds.

If an ideal counter has M counts in 1 seconds then

$$\begin{aligned} N &= M - MNt \\ N &= M(1 - Nt) \end{aligned}$$

If M_1 is the count rate with only source 1 present, M_2 is the count rate with only source 2 present and M_3 is the count rate with both sources 1 and 2 present then:

$$\begin{aligned} M_3 &= M_1 + M_2 \\ \frac{N_3}{(1 - N_2 t)} &= \frac{N_1}{(1 - N_1 t)} + \frac{N_2}{(1 - N_2 t)} \end{aligned}$$

Simple rearranging yields a quadratic in t:

$$0 = N_1 N_2 N_3 t^2 - 2N_1 N_2 t + (N_2 + N_1 - N_3)$$

The quadratic formula gives two results. A simple approximation to the quadratic in t can be made by assuming that since t is small $t^2 \approx 0$. Thus giving the equation:

$$N_2 + N_1 - N_3 = 2N_1 N_2 t$$

The solution to these two equations in can be compared, and it can be determined that the minus version of the quadratic formula gives the correct dead time.

8.2 Exponential Attenuation Results

lead sheet	none	q	r	s	t
number of counts	293	225	180	174	124
	250	227	172	177	146
	315	196	195	167	151
	305	196	198	162	137
	264	235	177	162	128
	287	223	200	182	138
	302	205	195	175	128
	275	211	213	160	120
	290	193	214	143	125
	315	214	185	145	123
count rate (s^{-1})	28.473	20.763	18.803	15.983	12.713
actual count rate (s^{-1})	28.894	20.986	18.985	16.115	12.796
mass lead ($mgcm^{-2}$)	0	1230	1890	3632	7435
thickness (cm)	0	0.081	0.163	0.318	0.635
kgm^{-2}	0	12.3	18.9	36.32	74.35
$(\ln M_0 - \ln M)^2$	0	0.0997	0.172	0.333	0.650

lead sheet	qt	rs	qs	st	rt
number of counts	139	134	152	109	99
	109	128	166	94	102
	150	156	181	91	100
	121	155	168	92	110
	113	141	149	104	128
	100	171	176	77	109
	117	140	149	110	111
	132	140	176	89	109
	135	145	165	92	98
	105	151	194	94	98
count rate (s^{-1})	11.7226	14.123	16.272	9.033	10.152
actual count rate (s^{-1})	11.793	14.225	16.409	9.075	10.206
mass lead ($mgcm^{-2}$)	8665	5522	4862	11067	9325
thickness (cm)	0.716	0.480	0.399	0.953	0.797
kgm^{-2}	86.65	55.22	48.62	110.7	93.25
$(\ln M_0 - \ln M)^2$	0.788	0.492	0.313	1.32	1.06

lead sheet	qr	qq	rr	ss
number of counts	204	205	170	134
	183	181	154	126
	195	221	163	130
	184	197	180	130
	164	201	162	127
	185	213	171	124
	161	188	172	141
	179	208	173	148
	190	211	154	150
	174	199	175	115
count rate (s^{-1})	17.703	19.752	16.253	12.763
actual count rate (s^{-1})	17.865	19.954	16.389	12.847
mass lead ($mgcm^{-2}$)	3120	2460	3780	7264
thickness (cm)	0.244	0.163	0.325	0.635
kgm^{-2}	31.2	24.6	37.8	72.6
$(\ln M_0 - \ln M)^2$	0.226	0.134	0.314	0.644

lead sheet	tt	rst	rsq	qst
number of counts	76	105	166	89
	74	105	128	95
	75	98	144	98
	79	85	163	94
	58	85	140	91
	100	105	140	91
	91	87	133	104
	72	86	133	102
	76	98	134	99
	83	92	139	100
count rate (s^{-1})	7.535	8.973	13.713	9.143
actual count rate (s^{-1})	7.381	9.014	13.810	9.186
mass lead ($mgcm^{-2}$)	14870	12957	6752	12297
thickness (cm)	1.27	1.12	0.561	1.04
kgm^{-2}	148.7	129.6	67.52	122.97
$(\ln M_0 - \ln M)^2$	1.83	1.33	0.534	1.29

lead sheet	qrt	rrs	rrt	rrq
number of counts	89	122	115	157
	108	143	110	144
	107	144	111	170
	113	123	100	164
	98	139	98	146
	117	136	98	174
	92	146	101	171
	105	112	117	164
	106	125	90	125
	93	133	103	139
count rate (s^{-1})	9.793	12.743	9.943	15.053
actual count rate (s^{-1})	9.842	12.826	9.993	15.170
mass lead ($mgcm^{-2}$)	10555	7412	11215	5010
thickness (cm)	0.879	0.643	0.960	0.406
kgm^{-2}	105.6	74.12	112.2	50.1
$(\ln M_0 - \ln M)^2$	1.14	0.646	1.11	0.406

6.2.3 Example Report: Latent Heat of Vaporisation of Liquid Nitrogen

Abstract

The latent heat of vaporisation of liquid nitrogen was calculated by measuring the boil-off from an insulated dewar flask when a known amount of heat was supplied. It was found to be significantly higher than the accepted value of 199 $kJkg^{-1}$ at 249 $kJkg^{-1}$. The high value can be accounted for because the volume measuring was stopped as soon as the heater was switched off, and not continued as the heater cooled. The specific heat capacity of a range of materials was then calculated over two temperature ranges, 293 K to 77 K and 87 K to 77 K. The results, though high in comparison with the accepted values were explainable by the high latent heat of vaporisation obtained in the first part of the experiment, are qualitatively consistent with Debye theory i.e. as the temperature drops the specific heat capacity drops too.

1 Introduction

1.1 Latent Heat of Vaporisation

The latent heat of vaporisation is the energy required to change the state of a material from a liquid to a gas at constant temperature.

1.2 Specific Heat Capacity

The specific heat capacity of a material is the energy required to raise the temperature of one mole of the material by one Kelvin.

1.3 History of Latent Heat

1.3.1 Joseph Black (1728 – 1799)

While at the University of Glasgow, Black was performing research on the freezing and melting of water and water/alcohol mixtures that led to the concept of latent heat of fusion. Similar work he performed later led him to establish the concept of latent heat of vaporisation – eventually leading to the general concept of specific heat capacity (c.1760).

1.3.2 Sir James Dewar (1842 – 1923)

While working on cryogenics at the Royal Institute between 1877 and 1904 Dewar invented the Dewar flask. The basic construction of the Dewar flask consists of a double walled glass/metal container with the gap between the walls 'filled' with a vacuum. In 1878 he achieved the liquefication of oxygen (boiling point 97 K). In the closing years of the 19th century he liquefied hydrogen (boiling point 20 K). His pioneering work on low temperature physics and vacuum techniques paved the way for a wide range of experimental work at very low temperatures. From 1904 to 1914 he extended his work to include low temperature calorimetry investigations.

1.4 History of Specific Heat Capacity

1.4.1 Dulong and Petit

Dulong and Petit developed the first theory behind specific heat capacity in 1819. This treated the atoms in a crystal as classical oscillators with $kT/2$ of energy per degree of freedom. This theory predicted the specific heat capacity to be $3R$ per mole. Although this is quite accurate at room temperature for many materials, it fails badly at low temperatures and even fails for some materials (e.g. diamond) at room temperature.

1.4.2 Einstein

Einstein observed that the specific heat capacity of a material changes with temperature and that not all materials had the same specific heat capacity at room temperature as Dulong and Petit had predicted. To explain this he developed a theory in which he viewed the atoms in a crystal as independent quantum harmonic oscillators, oscillating at one fixed frequency. This approach, though much more accurate than the Dulong and Petit method still failed at very low temperatures. Experimental results showed the dependence of specific heat capacity on temperature was T^3 rather than exponential as predicted by the Einstein approach.

1.4.3 Debye

The modern day theory was developed by Debye, who extended Einstein's approach by assuming that the oscillators were coupled and could vibrate at a whole range of different frequencies.

1.5 The Importance of Latent Heat and Specific Heat Capacity

The knowledge of the latent heat of vaporisation of material allows many calculations to be performed. The Clausius-Clapeyron equation is just one such equation that makes use of latent heat. Latent heat can also be used to calculate how much energy is needed to raise the temperature of a material from T_1 to T_2 when a change of state occurs between the two temperatures. The development of a theory of specific heat capacity has given a detailed understanding of how atoms in a crystal behave as temperature changes and led to the most appropriate material being used in many cases where the thermal properties of a material are an important consideration.

1.6 Aims of this Experiment

This experiment is designed to calculate the latent heat of vaporisation of liquid nitrogen and the specific heat capacity of a range of samples over two different temperature ranges. First, from 293 K (room temperature) to 77 K (boiling point of liquid nitrogen) and secondly over the range 87 K (boiling point of liquid argon) to 77 K. It also aims to qualitatively test the Debye theory that says that specific heat capacity decreases as temperature decreases.

2 Theory

2.1 Latent Heat of Vaporisation

The latent heat of vaporisation of a material is the energy required to change the phase of the material from a liquid to a gas at a constant temperature. In order for the phase change to occur energy must be supplied to the system. The energy taken in during the phase change does not go to increase the temperature of the material. Instead, this energy is used to provide the increased potential energy of the gas phase (i.e. overcoming attractive forces between the molecules) and, when the phase change results in expansion, to do external work in pushing back the atmosphere.

$$L_{vap} = \frac{Q}{m}$$

where L_{vap} is the latent heat of vaporisation, Q is the energy input and m is the mass of the sample that changes phase. The units used for are kJkg^{-1}.

2.2 Heat Capacity

The heat capacity, C_T, is defined as the ratio of the energy input by heating and the resulting temperature change (where ΔQ is the energy input and ΔT is the temperature change):

$$C_T = \frac{\Delta Q}{\Delta T}$$

In the limit of small ΔQ and ΔT, this becomes:

$$C_T = \frac{dQ}{dT}$$

This is an extensive parameter, dependent on the size of the sample. The units used for C_T are JK^{-1}.

2.3 Specific Heat Capacity

The specific heat capacity of a material is an intensive parameter, not dependent on the size of the system:

$$C = \frac{C_T}{n}$$

where n is the number of moles of material. The units used for are $JK^{-1}mol^{-1}$.

2.4 Debye Theory

The classical theory explaining the specific heat capacity of a solid assumes that in a crystal lattice there can only be vibrational energy – there is no rotation or translational energy. Each atom on a lattice site acts as an oscillator in three independent directions, so there are 3N oscillators (where N is the number of atoms in the lattice) each with two degrees of freedom (potential energy and kinetic energy). Classically there is $kT/2$ of energy per degree of freedom (where k is the Boltzman constant and T is the temperature) so:

$$(3N \times 2) \times \frac{1}{2}kT = 3NkT$$

From section 2.3

$$C_V = \frac{dQ}{dT} = 3R$$

where C_V is the specific heat capacity per mole at constant volume and R is the gas constant. This gives the Dulong and Petit value for the specific heat capacity. However, C_V is dependent on temperature, and clearly not independent as Figure 2.1 suggests:

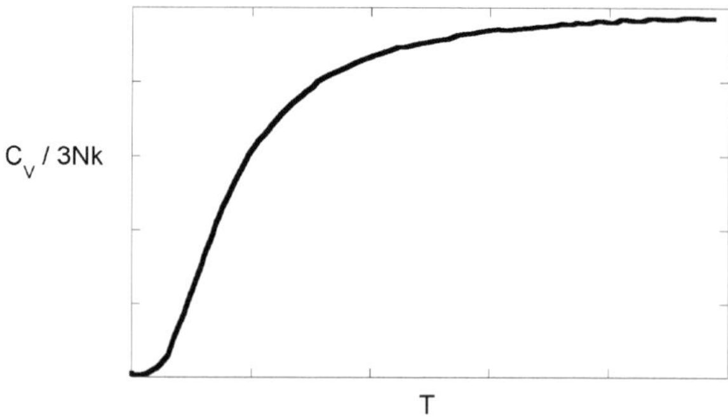

Figure 2.1: Graph to show the variation in C_V with temperature, T.

In an improved approach Einstein viewed the lattice as an array of independent quantum harmonic oscillators with fundamental frequency f. The Einstein temperature, where h is Plank's constant and θ_E is the Einstein temperature, is defined as:

$$\theta_E = \frac{hf}{k}$$

and

$$C_V = 3Nk\left(\frac{\theta_E}{T}\right)^2 \frac{e^{\theta_E/T}}{\left(e^{\theta_E/T} - 1\right)^2}$$

At high temperatures this gives the classical result, but at low temperatures it gives:

$$C_V = 3Nk\left(\frac{\theta_E}{T}\right)^2 e^{-\theta_E/T}$$

The Einstein approach gives a much better fit to experimental data than the classical (Dulong and Petit) approach, but there is still significant error, especially at low temperatures where $C_V \propto T^3$ and for metals at high temperatures. Einstein assumed that the oscillators are independent of each other; clearly this can't really be the case if the atoms are locked in a lattice structure. He also failed to take into account any directionality in the lattice structure and assumed there was only one possible vibrational frequency.

Debye recognised that the oscillators are not independent, but coupled, so that there are a range of vibrational frequencies corresponding to standing waves in the crystal structure. The disturbances can travel around the material in any direction.

The density of available states (or frequencies) is:

$$D_{Debye} = \frac{12\pi V f^3}{c_s^3}$$

where V is volume and c_s is the speed of sound. Since the most vibrations which can occur is $3N$, the maximum number of states is also $3N$:

$$\int_0^{f_{max}} \frac{12\pi V f^2}{c_s^3} df = 3N$$

so:

$$f_{max} = \sqrt[3]{\frac{3Nc_s^3}{4\pi V}}$$

The maximum frequency defines a characteristic temperature, the Debye temperature, θ_D where $a = (V/N)^{1/3}$ is the average interatomic spacing:

$$k\theta_D = hf_{max}$$

$$\theta_D = \left(\frac{3}{4\pi}\right)^{1/3} \frac{hc_s}{ka}$$

Quantum theory restricts the energy of a sound wave to increments of hf. With each extra hf of energy it is said that an extra phonon (the quantum of elastic energy in a lattice) is present. The estimated number of phonons in the mode with frequency f is:

$$\overline{n}(f) = \frac{1}{e^{hf/kT} - 1}$$

The energy present in the lattice is then:

$$\langle E \rangle = \int_0^{f_{max}} hf D_{Debye} \overline{n}(f) df$$

$$\langle E \rangle = \int_0^{f_{max}} \frac{12\pi hV}{c_s^3} \frac{f^3}{e^{hf/kT} - 1} df$$

At low temperatures, $T \ll \theta_D$, this becomes:

$$\langle E \rangle = \frac{12\pi h V}{c_s^3}\left(\frac{kT}{h}\right)^4 \times \int_0^{\theta_D/T} \frac{x^3}{e^x - 1} dx$$

where $x = hf/kT$. So:

$$C_V = \left(\frac{\partial \langle E \rangle}{\partial T}\right)_V = \frac{12\pi^4}{5} Nk \left(\frac{T}{\theta_D}\right)^3$$

At high temperatures $T \gg \theta_D$ it becomes $3NkT$ and $C_V = 3R$ per mole.

C_P (specific heat capacity at constant pressure) is the most appropriate quantity to use in this experiment since atmospheric pressure will remain constant. But since the solids tested here contract only very little compared to gases there is no need to draw a distinction between C_P and C_V.

2.5 Energy Dissipated by a Resistor

By definition, the potential difference, V, between two points is equal to the work done, W, in moving a positive charge, Q, from the lower potential to the higher potential. Thus:

$$W = QV$$

If the current, I, is constant (where t is time):

$$Q = It$$

So:

$$W = It(V_2 - V_1)$$

If $V_1 = 0$, then the energy dissipated by a resistor (i.e. a heater) is:

$$W = VIt$$

2.6 Number of Moles of a Gas

The gas volume meter records the number of litres of gas produced. In order to perform calculations using this result the volume is needed in terms of numbers of moles, not number of litres.

Using the equation of state of an ideal gas:

$$PV = nRT$$

where P is pressure, V is volume, n is number of moles, R = 8.31 $JK^{-1}mol^{-1}$ is the gas constant and T is the temperature, the volume of one mole of gas at room temperature (20°C) and pressure (101325 Pa) is found to be 24 dm^3. (This theory assumes that the nitrogen gas has reached room temperature by the time it has traveled through a long tube and goes into the gas volume meter.)

2.7 Error Calculations

Errors were calculated on all results using the formulae given in Lyons [7] for the propagation of errors. If:

$$a = b \pm c$$

then

$$\sigma_a^2 = \sigma_b^2 + \sigma_c^2$$

and if:

$$f = xy \text{ or } f = \frac{x}{y}$$

then:

$$\left(\frac{\sigma_f}{f}\right)^2 = \left(\frac{\sigma_x}{x}\right)^2 + \left(\frac{\sigma_y}{y}\right)^2$$

where σ_k is the standard deviation of the variable k.

3 Experimental Methods

3.1 Latent Heat of Vaporisation of Liquid Nitrogen

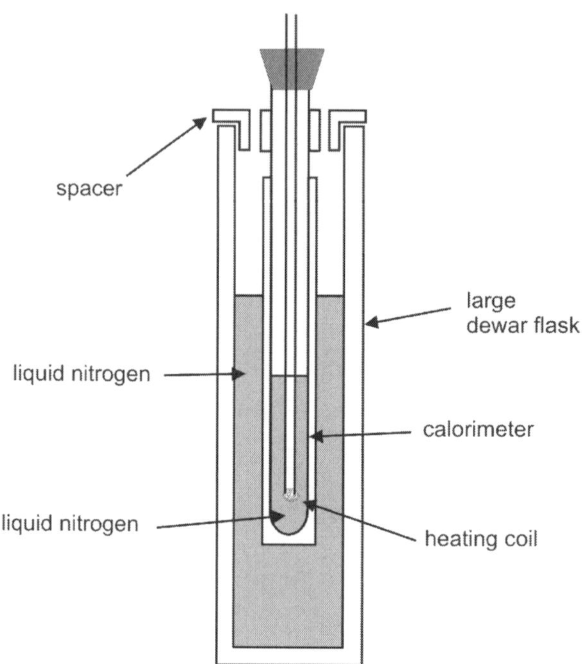

Figure 3.1 – The Latent Heat Calorimeter. A diagram of the apparatus used in section 3.1.

In order to calculate the latent heat of vaporisation of liquid nitrogen a large dewar flask was filled with liquid nitrogen. The calorimeter was carefully inserted into the large dewar flask and then also filled with liquid nitrogen. The heating coil was then inserted into the calorimeter. The apparatus was allowed to cool down. During the initial cooling of the apparatus both the large dewar flask and the calorimeter were topped up with liquid nitrogen as the boil off was considerable. When the boil off had reduced to a steady and low rate, the experiment was started.

Firstly a background rate was taken, so as this contribution could be subtracted from the results so that the contribution from the heating coil alone could be calculated.

The heating coil was then turned on, and the current and (electric) potential difference were recorded using a multimeter. As the liquid nitrogen in the calorimeter boiled off the volume was recorded from the gas volume meter.

Using the current and (electric) potential difference supplied to the heater and the time period over which the measurements were taken, the heat dissipated by the heater was calculated. The volume of N_2 evolved as a result of this amount of heat was read from the gas volume meter. This was converted into the number of moles as by dividing by 24 (as outlined in section 2.6).

The mass of N_2 produced was calculated using:

$$n = \frac{m}{M_r}$$

where n is the number of moles, m is the mass in grams and M_r is the atomic mass in grams (28 in the case of N_2). L_{vap} was then calculated.

3.2 Specific Heat Capacity Over Temperature Range 293 K → 77 K

The heating coil was replaced by a stainless steel tube with a bladder and rubber hose attached. The bladder is needed to help prevent air condensing inside the tube. Air is mainly made up of nitrogen and oxygen. In the stainless steel tube submerged in liquid nitrogen, oxygen can condense since its boiling point is 97 K and that of nitrogen is 77 K. As well as producing undesired inaccuracies in the experiment, having liquid oxygen present is a fire hazard.

The most effective way to prevent oxygen condensing is for it to not be present at all in the stainless steel tube. This can be achieved by creating an atmosphere of nitrogen inside the tube. Firstly the tube was partially cooled in the calorimeter so that the next part could be performed more easily. The tube was then removed from the calorimeter and a little liquid nitrogen was added to it. The end of the rubber hose was corked so as to seal the tube. As the liquid nitrogen boiled off, the bladder enlarged. While the tube and hose were held upside down with the cork removed the bladder was squeezed and the nitrogen gas expelled. This was repeated until all the nitrogen had boiled off (i.e. the bladder stopped enlarging when the hose was re-corked). Since oxygen gas ($1.43\,\text{kgm}^{-3}$) is denser than nitrogen gas ($1.25\,\text{kgm}^{-3}$), is sinks out of the tube, encouraged by the nitrogen, thus leaving an oxygen-free atmosphere inside the stainless steel tube. The hose was then immediately re-corked. This method proved very successful at preventing the formation of liquid oxygen in the stainless steel tube.

The stainless steel tube was then reinserted into the calorimeter and any significant boil off from the liquid nitrogen in the calorimeter or the large dewar was topped up. When the boil off had settled to a low steady value a background reading was started.

After the background reading was taken the cork was briefly removed and the sample was added to the hose while squeezing the bladder so as no air could enter the tube or hose. The hose was then immediately re-corked. The sample dropped into the bottom of the stainless steel tube and readings were taken from the gas volume meter every 30 seconds until the rate of boil off had dropped below that of the background rate.

The sample was not added to the hose earlier and left there while the background reading was taken (with the hose bent over to prevent the sample entering the calorimeter) since the ambient air temperature inside the hose was not the same as the ambient air temperature in the room. This would have meant the temperature range over which the measurements were taken would not have been known.

Each time a reading was taken a new background reading was taken since it was more accurate than trying to keep the level of the liquid nitrogen in the calorimeter and the dewar constant between readings.

This part of the experiment was performed with samples of copper, aluminium, graphite, zinc, lead and rock salt.

3.3 Specific Heat Capacity Over Temperature Range 87 K $\rightarrow$ 77 K

The same procedure was followed as in section (3.2) but this time the samples were first cooled in liquid argon (boiling point 87 K) before being inserted into the stainless steel tube. The liquid argon was made by passing argon gas through a copper tube submerged in liquid nitrogen. The samples were attached to a piece of thread so as they could be removed from the liquid argon quickly and easily and inserted in to the stainless steel tube. In the light of the previous work an improved method for taking a background reading was introduced. This consisted of reading the gas volume meter every 30 seconds for 600 seconds and then inserting the cooled sample and taking another set of readings every 30 seconds for 600 seconds. This provided a more accurate representation of the background reading than earlier where it was assumed to be constant. Clearly, the calorimeter can't be perfectly insulated, especially with a large conducting metal tube sticking out into the air, so the background reading can't be exactly constant.

3.4 Volume and Density

All the samples were weighed using an accurate balance. The volume of the samples was calculated using a measuring cylinder half full of distilled water. The volume was recorded before and after the samples were placed in the water. The volume of the sample is then the difference in these volumes. The density was then calculated using the mass and volume of the sample.

3.5 Safety

While performing this experiment careful consideration was given to safety. Both the liquid nitrogen and liquid argon used are obviously very cold. Goggles were used to prevent splashes entering the eyes. Gloves were also worn when handling the apparatus as it was very cold. The liquids were added very slowly and carefully to the apparatus (which was at room temperature initially) to prevent violent boil off. It was also noted that the extreme cold causes glassware to become brittle, so glass-glass and glass-metal contact was avoided as much as possible. The spacer holding the calorimeter in the large dewar has a loose fit to allow any excess gas to escape so as pressure does not build up. The possible formation of liquid oxygen (highly flammable) was avoided by using the method outlined in section 3.2.

3.6 Calculation of Results Over Temperature Range 293 K → 77 K

The volume of N_2 evolved over the previous 30 seconds was calculated from the results. This was then plotted against time to produce a graph showing the rate of N_2 boil-off (see Figure (4.1) for an example of this graph). The background reading was then added to the same graph. At the point where the two lines meet, it is assumed that the sample is at 77 K (the same temperature as the liquid nitrogen). A line of best fit was added to this graph using a spreadsheet program to enable the intercept (the time when the temperature of the sample reached 77 K) to be calculated exactly.

The original results from the gas volume meter were then plotted against time, and a line of best fit added using the spreadsheet program (see Figure (4.2) for an example of this graph). When the intercept from the first graph was substituted into the equation of the second line of best fit, the reading on the gas volume meter at the time when the temperature of the sample was 77 K was found. The reading from the gas volume meter at $t = 0$ was then subtracted from this to find the actual volume of N_2 evolved.

This volume was then converted into the number of moles of N_2 using the theory laid out in section (2.6) and into a number of grams.

Using the average value of latent heat of vaporisation of N_2 calculated earlier this was converted into a number of Joules. Dividing by the change in temperature (i.e. $293\,\mathrm{K} - 77\,\mathrm{K} = 216\,\mathrm{K}$ or $87\,\mathrm{K} - 77\,\mathrm{K} =$

10 K) gave a figure in Joules per Kelvin. Using the mass of the sample and the Mr, the number of moles of the sample was calculated and then the specific heat capacity in Joules per Kelvin per mole ($JK^{-1}mol^{-1}$) was calculated.

3.7 Calculation of Results Over Temperature Range 86 K $\rightarrow$ 77 K

The background reading observed during the first 600 seconds was extrapolated to cover the time period of the rest of the experiment. The difference between the actual volume and the extrapolated volume was then used in the same way as the change in volume was used in section 3.6.

4 Observations and Results

4.1 Latent Heat of Vaporisation Results

	Volume $N_2(\text{dm}^3)$	Time (s)		Volume $N_2(\text{dm}^3)$	Time (s)
Reading 1			Reading 2		
Background	1.3	1469	Background	10.5	600
Reading	93.6	1200	Reading	53.5	600
Current (A)	2.4		Current (A)	2.4	
Voltage (V)	8.59		Voltage (V)	8.53	
$L_{vap}(\text{kJkg}^{-1})$	229		$L_{vap}(\text{kJkg}^{-1})$	246	
Reading 3			Reading 4		
Background	1.8	600	Background	3.6	600
Reading	17.2	600	Reading	80.3	600
Current (A)	1.0		Current (A)	3.2	
Voltage (V)	3.63		Voltage (V)	11.54	
$L_{vap}(\text{kJkg}^{-1})$	275		$L_{vap}(\text{kJkg}^{-1})$	248	

Table 4.1 – Latent heat of vaporisation results

The average L_{vap} was found to be $249\,\text{kJkg}^{-1}$ with a propagated error calculated to be $\pm 4\,\text{kJkg}^{-1}$.

4.2 Mass, Mr, Volume and Density of Samples

Sample	Mass (g)	Mr	Volume (cm^3)	Density (kgm^{-3})
Al	22.6541	27	8.5	2670 ± 20
Cu	48.1326	63.5	5.5	8750 ± 80
Zn	40.2592	65.5	5.5	7320 ± 70
C	10.7952	12	6	1800 ± 20
Pb	71.0565	207	6.5	10930 ± 90
NaCl	5.3496	58.5		

Table 4.2 – Sample measurements

4.3 Specific Heat Capacity Over Temperature Range 298 K $\rightarrow$ 77 K

Sample	Specific Heat Capacity ($JK^{-1}mol^{-1}$)
Al	30.5 ± 0.5
Cu (1st Result)	24.0 ± 0.4
Cu (2nd Result)	39.6 ± 0.7
Zn	22.5 ± 0.4
C (1st Result)	15.6 ± 0.3
C (2nd Result)	8.1 ± 0.2
Pb	30.3 ± 0.6
NaCl	45.5 ± 1.6

Table 4.3 – Specific heat capacity results

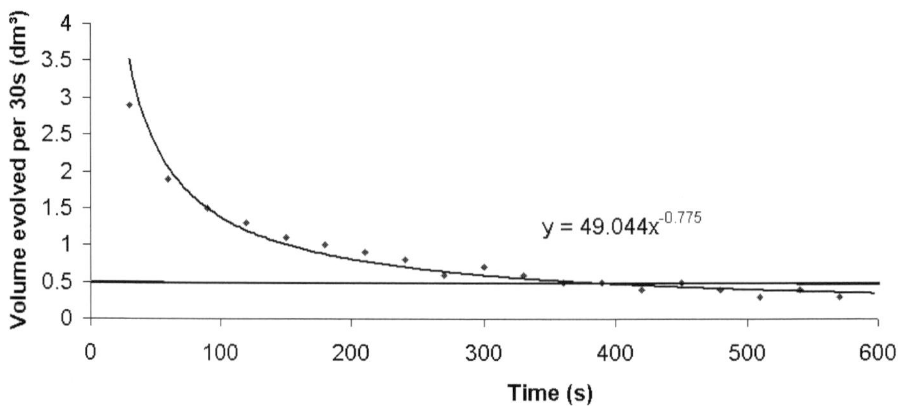

Figure 4.1 – Example of a graph used to calculate the time when the background level was reached. This graph refers to the first result for copper over the range 293 K to 77 K. The horizontal line shows the background rate and the curve is a line of best fit added to the points showing the rate at which the nitrogen was evolved during the experiment. The equation is the equation of the line of best fit.

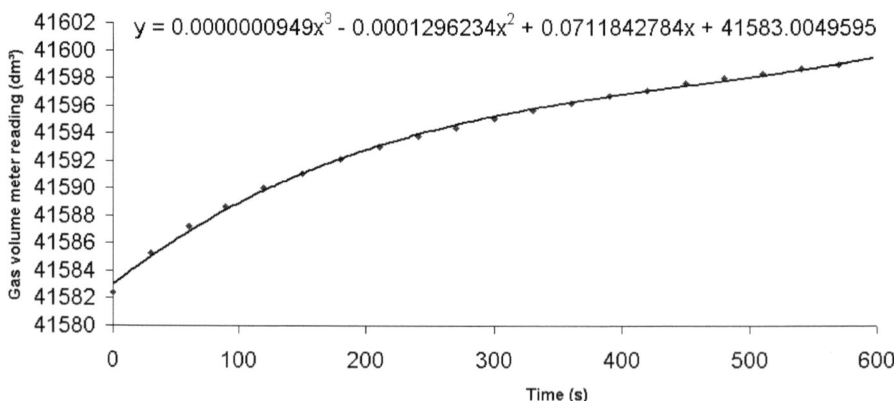

Figure 4.2 – Example of a graph used to calculate the volume of nitrogen evolved in the time taken for the rate of production of nitrogen to reach background levels. This graph again refers to the first result for copper over the temperature range 293 K to 77 K. The curve is the line of best fit added to the points showing the reading on the gas volume meter. The equation is the equation of the line of best fit.

Note on validity of equations of lines of best fit: the choice of the form of these equations was based upon finding an equation that fitted the points well and not upon any assumed physical dependence of one variable on another. This method is merely an improved version of drawing a straight line up from the x-axis to a point on a hand-drawn line of best fit and the drawing a line across to the y-axis to take a result. Clearly the equations of the lines of best fit are not valid outside the range over which the experiment was performed, but within the range they provide the most accurate method of evaluating results from the graphs.

4.4 Specific Heat Capacity Over Temperature Range 87 K → 77 K

Sample	Specific Heat Capacity ($JK^{-1}mol^{-1}$)
Cu	11.5 ± 3.8
C	3.23 ± 3.0

Table 4.4 – Specific heat capacity results

4.5 Accepted Results

4.5.1 Latent Heat of Vaporisation

The accepted value of the latent heat of vaporisation of N_2, L_{vap}, is 199 $kJkg^{-1}$.

4.5.2 Specific Heat Capacity

Material	Specific Heat Capacity ($JK^{-1}mol^{-1}$) at Temperature (K)					
	50 K	100 K	150 K	200 K	298 K	400 K
Al	3.8	13.1	18.5	21.6	24.4	25.6
Cu	6.3	16.2	20.5	22.7	24.4	25.1
Zn	11.1	19.2	22.5	23.8	25.4	26.4
C	0.5	1.7	3.2	5.0	8.5	11.9
Pb	21.4	24.4	25.4	25.9	26.4	27.4
NaCl					49.7	
N_2	29.1	29.1	29.1	29.1	29.2	

Table 4.5.1 – Specific heat capacity results

Density

Material	Density (kgm^{-3}) at 298 K)
Al	2698
Cu	8933
Zn	7135
C	2266
Pb	11343
NaCl	2165
N_2(gas)	1.25
O_2(gas)	1.43

Table 4.5.2 – Table of densities

5 Discussion

5.1 Latent Heat of Vaporisation of N_2

It was found that $L_{vap} = 249 \pm 4\,\mathrm{kJkg^{-1}}$. In comparison with the accepted value of 199 $\mathrm{kJkg^{-1}}$, the experimental value appears to have some significant non-statistical error since the accepted value lies well outside the statistical error range.

By examining the experimental results it can be seen that 'too much energy is needed to boil off each kg of nitrogen'. The energy is probably being lost in the heater. When the power supply to the heater was turned off the volume of nitrogen evolved was no longer measured. At this point some of the energy calculated to have been transferred to the liquid nitrogen had not done so since the heater was sill hot.

The experiment could be modified so as volume readings were continued after the heater was turned off until the rate of production of the gas reduced to background levels. This way all the energy calculated to have gone into the heater would have been transferred to the liquid nitrogen. This should produce a more accurate value for the latent heat of vaporisation.

5.2 Specific Heat Capacity Over Temperature Range 298 K → 77 K

The results calculated for the aluminium, the second copper reading, the first carbon reading and the lead readings were all significantly too high when compared with the accepted results. (See section 4.3 and 4.52). The accepted results give the specific heat capacity at a particular temperature and are not directly comparable to those obtained experimentally over a large temperature range. But, since specific heat capacity is a single valued function of T which increases as temperature increases, it is expected that the specific heat capacity over a large temperature range should fall between the accepted specific heat capacities at the bottom and the top of the temperature range. For these results this clearly isn't the case.

The other results (i.e. the first copper, the zinc, the second carbon and the salt) fall within the accepted results given for 50 K and 298 K – this is to be expected since the experiment was carried out between these temperatures.

The fact that the experimental results are consistently too high in comparison with the accepted values is explained by the latent heat of vaporisation for liquid nitrogen calculated earlier in the experiment being significantly too high. If this had been more accurate then it would have reduced the results for specific heat capacity, making them in much better agreement with the accepted values.

Another possible factor contributing the high values obtained for the specific heat capacities was the variation in the background count between the beginning and end of the experiment. It was assumed to be constant when testing over the range 293 K → 77 K. This inaccuracy was rectified by an improved method for testing over the range 87 K → 77 K as outlined in section 3.3.

5.3 Specific Heat Capacity Over Temperature Range 87 K → 77 K

The general trend shown by the Debye model that the heat capacity decreases as temperature decreases can be observed from these results in comparison with those obtained over the temperature range from 293 K to 77 K. The specific heat capacity over the range 293 K to 77 K is much higher than that over the range 87 K to 77 K.

The results are still too high in comparison with the accepted results. The error range is very high, and nearly as large as the result itself for carbon, since the volume of nitrogen evolved is very small. (Since the sample is at 87 K when it is put into the liquid nitrogen at 77 K, there is much less heat energy that can be used to vaporise the liquid nitrogen than when the sample is originally at 293 K). Again the reason for the results being too high is the very high latent heat of vaporisation obtained earlier.

5.4 Density The density of the samples was calculated, so as this could be compared to the accepted density of the pure material (for which the accepted results are quoted). There is a small difference between the calculated and accepted densities, but this should not be significant enough to account for the variation seen between the experimental and accepted values of the specific heat capacity.

6 Conclusion

The latent heat of vaporisation of liquid nitrogen was found to be $249 \pm 4\,\mathrm{kJkg^{-1}}$, significantly higher than the accepted value of $199\,\mathrm{kJkg^{-1}}$. The specific heat capacity over the temperature range from 293 K to 77 K was generally found to be higher than the accepted value, but this is explained by the latent heat of vaporisation being significantly higher than the accepted value. When the samples were cooled to liquid argon temperature, the specific heat capacities still remain high, but at one extreme of the error range the results are comparable with accepted values.

As the temperature range drops, the specific heat capacity decreases qualitatively inline with Debye theory and the accepted values.

7 Bibliography

[1] MacDonald, D. K. C. (1964) Introductory Statistical Mechanics for Physicists, John Wiley and Sons (USA)

[2] Reif, F (1965) Statistical and Thermal Physics, McGraw-Hill Book Company (London)

[3] Baierlein, Ralph (1999) Thermal Physics, Cambridge University Press (Cambridge)

[4] Website: Archives in London and the M25 Area, www.aim25.ac.uk

[5] Website: Glasgow University Chemistry Department, http://www.chem.gla.ac.uk/ alanc/dept/black.htm

[6] Muncaster, Roger (1993) Physics, Stanley Thornes (Publishers) Ltd. (Cheltenham)

[7] Lyons, Louis (1991) A Practical Guide to Data Analysis for Physical Science Students, Cambridge University Press (Cambridge)

[8] Kaye and Laby (1973) Tables of Physical and Chemical Constants, Longman

[9] Dewar, J. Proc. Royal Society A76 325, (1905)

References

[1] http://www.cleapss.org.uk.

[2] Mythbusters TV Programme, Discovery Channel, Episode 63 (2006).

[3] British Compressed Gases Association Code of Practice CP30 'The Safe Use of Liquid Nitrogen Dewars up to 50 litres'.

[4] http://www.esrl.noaa.gov/gmd/ccgg/trends/global.html.

[5] Bureau International des Poids et Mesures: www.bipm.org/utils/common/pdf/its-90/SUPChapter4.pdf.

[6] M. M. J. French and Michael Hibbert. Making liquid oxygen. *Phys. Ed.*, 45:221, 2010.

[7] M. M. J. French. Mobile phone faraday cage. *Phys. Ed.*, 46:290, 2011.

[8] J. G. Ziegler and N. B. Nichols. Optimum settings for automatic controllers. *Transactions of the ASME.*, 64:759, 1942.

[9] A. S. McCormack and K. R. Godfrey. Rule-based autotuning based on frequency domain identification. *IEEE Transactions on Control Systems Technology*, 6:1, 1998.

[10] http://www.amsterchem.com/onlinehelp.php?page=welcome.htm.

[11] M. Vannoni and S. Straulino. Low-cost accelerometers for physics experiments. *Eur. J. Phys.*, 28:781, 2007.

[12] F. Rooney and W. Somers. Using the wiimote in introductory physics experiments. *TCNJ J. Stud. Scholarsh.*, 12:1, 2010.

[13] M. D. Wheeler. Physics experiments with nintendo wii controllers. *Phys. Ed.*, 46:57, 2011.

[14] http://wiimotephysics.codeplex.com/.

[15] http://pdg.lbl.gov/2012/tables/rpp2012-sum-quarks.pdf.

[16] J. A. Nelder and R. Mead. A simplex method for function minimisation. *Computer Journal*, 7:308, 1965.

[17] http://www.openoffice.org.

[18] http://www.miktex.org.

[19] http://www.winedt.com.

[20] http://www.irfanview.com/.

[21] M. M. J. French et al. Changing fonts in education: How the benefits vary with ability and dyslexia. *The Journal of Educational Research*, 2013.

[22] http://www.prezi.com.

Index